Eleanor Ormerod

A manual of injurious insects with methods of prevention and remedy for their attacks to food crops

Eleanor Ormerod

A manual of injurious insects with methods of prevention and remedy for their attacks to food crops

ISBN/EAN: 9783337201326

Printed in Europe, USA, Canada, Australia, Japan

Cover: Foto ©berggeist007 / pixelio.de

More available books at **www.hansebooks.com**

A

MANUAL OF INJURIOUS INSECTS,

WITH METHODS OF

PREVENTION AND REMEDY

FOR THEIR ATTACKS TO

FOOD CROPS, FOREST TREES, AND FRUIT,

AND WITH

Short Introduction to Entomology.

BY

ELEANOR A. ORMEROD, F.M.S., &c.,

AUTHOR OF THE 'COBHAM JOURNALS,' 'REPORTS ON INJURIOUS INSECTS,' ETC.

LONDON: W. SWAN SONNENSCHEIN & ALLEN, PATERNOSTER SQUARE.
EDINBURGH: J. MENZIES & CO., HANOVER STREET.

LONDON:
PRINTED BY WEST, NEWMAN AND CO..
HATTON GARDEN, E.C.

TO THE

LANDOWNERS,

FARMERS, FORESTERS, AND GARDENERS,

OF

GREAT BRITAIN AND IRELAND,

AND ESPECIALLY TO

THOSE WHO HAVE ASSISTED IN THE WORK,

This Volume

IS RESPECTFULLY INSCRIBED

BY

THEIR OBEDIENT SERVANT,

THE WRITER.

PREFACE.

BEFORE entering on this volume I wish to express the obligations I am under to the many contributors who have helped me to form it, and to thank them not only for the information which they have allowed me to publish, but also for the invariable kind courtesy with which my applications have been received.

A work of this kind, involving measures of treatment of the most varied nature, must necessarily be a compilation from many sources, and these I have acknowledged as far as possible throughout, either by appending initials, the name being given in full in the list of contributors; or by the title of the publication from which the information was taken, and by reference to which the various points may be verified. Some of the assistance, however, has been so continuous throughout the work that I cannot thus fully acknowledge it, and I wish especially to express my thanks to Mr. Malcolm Dunn, of Dalkeith, first, for aiding me in forming the plan of the volume, and for much valuable information and suggestion; as well as for the introductions, by means of which much excellent contribution has been obtained. To Mr. George Brown, jun., Watten, Caithness, I offer my thanks for advice and assistance throughout the progress of the work, and especially for his valuable help and information regarding points of agricultural treatment entered on in the part regarding "Food Crops," for which I am much indebted. To Mr. Charles Whitehead I am greatly obliged for information regarding insect-prevention in the Hopgrounds; and I desire also to acknowledge with thanks the courteous offers of assistance, and valuable donations of books on the subject of Economic Entomology received from various members of the Entomological Department of the Government of the United States of America.

I should wish to add the name of my sister and constant assistant, Miss G. Ormerod, to those who have aided me in this work.

Many of the illustrations are reproductions from the beautiful figures drawn from life by John Curtis for his 'Farm Insects,' for the use of which I am indebted to the kind courtesy of Messrs. Blackie & Son, Glasgow; and I am under obligation to the proprietors of the

'Gardeners' Chronicle' for extended permission to avail
myself in this volume of the electros of which use was
originally kindly allowed for my Reports; these are
partly from the pencil of John Curtis, partly from that
of Prof. Westwood. I am also indebted to Mr. John
Kirchner for the care and skill with which he has
engraved the figures placed in his hands for this
volume.

In offering this work I am only too well aware how
imperfect it is, partly from the need that exists of
more information regarding habits of insects (especially
forest pests), together with methods of prevention
recorded from observations in this country; partly
from my own inability to perform the task as such a
work should be done. I have earnestly endeavoured
to form a volume that might, happily, be of some
service in this important field; but to do this work
thoroughly requires great powers, and it is only
through the kind assistance afforded me that I have
been able to do anything towards it; and it is not
without uneasiness that I venture to lay my attempts
before readers well able to form a better judgment than
myself on some, perhaps many, of the points which
have been entered on.

Any information that may kindly be forwarded by those practically acquainted with the subject towards supplying points that are deficient would be gladly received, either for addition or correction, should the work pass to a second edition; or for insertion in the 'Yearly Reports on Injurious Insects.'

ELEANOR A. ORMEROD.

Dunster Lodge, near Isleworth,
   May 24th, 1881.

CONTENTS.

INTRODUCTION.

THE object of the present volume is to give in as concise a form as possible some account of the insects that are commonly injurious to our food crops, forest trees, and fruit, together with methods of treatment that have been found serviceable in preventing attacks, or in averting serious damage.

It is not possible, in the limits of a Manual intended to be really a "handy book," to offer anything like complete details of the life-histories of the insects, but it has been endeavoured to notice such of the more important points in their appearance, their methods of attack, and the various transformations they undergo in the different stages of their existence, as with the help of the figure may enable the reader to ascertain which of the insects that are commonly injurious his crop may be suffering from.

The different kinds of attack are arranged alphabetically under the headings of Food Crops, Forest Trees, and Fruit, beginning respectively with Asparagus, Ash, and Apple; and the insects attacking each crop or tree are similarly arranged alphabetically under such crop, by the name by which they are *commonly* known, as Turnip "Fly," "Wireworm," &c.

The number of synonyms or *scientific names* used by different writers cause some difficulty; therefore, in

cases of doubt, where the writer quoted from is one of
our known authorities, the name *used* by him has been
selected that the information might be preserved
entire.

With regard to methods of prevention : the most
serviceable of these are based, *not* on applications when
the crop or trees are undergoing attack, but rather on
modes of cultivation and treatment which may diminish
the amount of insect-presence beforehand, by clearing
away all points of harbourage and breeding-places, as
well as attacking the pest generally and on a broad scale
at the points where the details of its habits show that it
is most open to injury. Also on means adapted to
promote vigorous and healthy growth, whereby we lessen
the liability of the plant to sustain damage, first, by
pushing it forward in its early stages, and afterwards by
enabling it to overcome or outgrow the damage it may
suffer in cases of ordinary amount of attack.

The difficulty is when we get to special applications,
for in these cases the success of the remedy depends on
many circumstances, such as weather, character of the
soil, condition of the crop, and the time of day when
used; as many dressings applied when the dew is on
will totally fail if used when the leafage is dry or in the
heat of the day.

In the various methods of treatment mentioned
throughout the volume an endeavour has been made to
give only such as have been found to answer by trust-
worthy observers, or which by comparison of observations
may be presumed to be of service; but many points of
agricultural practice necessarily vary as much as the
soils do in different parts of our island, and the reader
is especially requested to bear this in mind where

treatment recorded as serviceable in one district should not be in accordance with what is desirable in another, as it is impossible in this little book to give the full details; and the best treatment—even such as has been worked out by men thoroughly acquainted with the science and practice of their calling—cannot be universally applicable.

In a very few instances treatment which must be altogether exceptional has been recorded, *attention being drawn* to this point. This is especially the case with the large amount of Turnip-seed noted as having been advised by some growers, at page 149, in order to make up such a thick plant as might counterbalance the ravages of the Turnip Fly; an amount by which, failing the insect-attack, the crop would be *ruined* by a few days' neglect in singling, and far exceeding anything serviceable on land which can be pulverized to a fine tilth, and where seed at the rate of two and a half to four pounds the acre is amply sufficient; but a very different state prevails on many clay soils, where mechanical means are often insufficient to reduce the land to a proper surface, and consequently much seed is wasted and the rough ground forms a most favourable shelter for the " Fly "; the suggestion, however, must be looked on as quite exceptional, and in two other cases of remedies advised theoretically, but very unsafe in practice, attention is drawn to the point by italics.

A few words may be added regarding the properties of gas-lime as a protection against insect-attack, which are not yet so well known as they ought to be. On its first removal from the purifying-chambers it is destructive of life to insect and vegetable alike, but by exposure to atmospheric action, or by being mixed with the soil, it absorbs oxygen, and gradually changes into sulphate of lime or gypsum—a safe and serviceable manure. It is a valuable dressing for insect-infested land, laid on, and dug or ploughed in, in the autumn. Notes of application of this remedy to special insects' attacks are given.

There are many other points which it might be well to enter on further, but space does not allow; and having been desired to add some notes which might serve as a kind of Introduction to Entomology, it is endeavoured in the following pages to give as shortly as possible some information regarding the main points of insect-life, such as the egg; the characteristics of insects in their three successive states of life; the principles of classification, with a table of the thirteen orders into which insects are classified; and a short description of the chief distinctions of these orders; to which (preceding the Index) a Glossary is added, with explanation of some of the most commonly used entomological terms.

INTRODUCTION TO ENTOMOLOGY.*

INSECTS begin their lives either by being hatched from eggs, or produced alive by the female; commonly they are hatched in the form known as maggots, caterpillars, or grubs, but they are never generated by decaying vegetables, putrid water, bones, carcases, dung, or any other matter, dead or alive, excepting their own insect forerunners. They come out of these matters constantly, but, if the observer will watch, he may often see the arrival of the insects, the laying of the eggs, and be able to satisfy himself as to the gradual development and the method of breeding, and that the progeny is produced by the female insect.

The eggs are usually laid soon after the pairing of the male and female, and are deposited on or near whatever may be the food of the larvæ. They are laid singly or in patches, and are sometimes attached by a gummy secretion to the leaf or whatever they are laid on; occasionally they are fastened by a short thread, or raised (like the heads of pins) on a stiff foot-stalk of hardened viscid matter. Such insects as insert their eggs in living animal or vegetable matter are furnished with a special egg-laying apparatus or *ovipositor*, such as a borer, or organs enclosing bristle-like points or saws, by means of which the female pierces a hole, and passes the egg down into the wounded spot.

For the most part insect-eggs hatch shortly after they are laid, but sometimes they remain unhatched during

---

* For the meaning of special entomological terms, and fuller explanations of some of the details, the reader is referred to the Glossary preceding the Index.

the winter; and it is believed that, where circumstances
are unfavourable to development, they may remain
unhatched for years, but this point is one of those
subjects on which more information is needed. They
have been found to endure intense cold without injury,
and, besides some special and extraordinary instances, it
has been found by experiment that insect-eggs may be
exposed to a temperature lower than that to which they
are usually subjected in this country, and cold enough
to solidify their contents without destroying their powers
of hatching.

In a very few cases insects are partly developed
before birth, otherwise, after hatching from the egg, or
being produced alive (in the same first stage of development) by the female, insects pass their lives in three
different conditions or stages successively.

The first is that in which they are known as maggots,
grubs, or caterpillars; in the case of Grasshoppers,
Cockroaches and some other insects, where the young
are very much the same shape as the parent, only
without wings, they usually go by the parent's name;
the young of Green-fly are sometimes known as "nits."
In this state they are active, voracious, and increase in
size; and in this first stage all insects are scientifically
termed *larvæ*.

In the second stage some orders of insects are usually
inactive and cannot feed, as is the case with the chrysalis
of the Butterfly, or the mummy-like form of the Beetle
or Wasp with its limbs in distinct sheaths folded down
beneath it; some, however, are active and feed, as
Grasshoppers, Cockroaches, Aphides (or Green-fly), and
others, and resemble the parent insect, excepting that
their wings and for the most part their wing-cases are
not as yet fully formed; and in this second stage all
insects are scientifically termed *pupæ*.

The third state is that of the perfect insect, in which
(whether male or female, or of whatever different kind,
as Moth, Beetle, Cricket, Aphis, &c.) it is scientifically
termed an *imago*.

The term *Larva* is from the Latin, meaning a mask or ghost, and signifies that the insect in this stage gives a mere vague idea of its perfect form.

Pupa signifies an infant, and is appropriate to the second stage in which the insect is *forming* into the perfect state, but is not fully developed either in its limbs or functions.

Imago signifies the image, the likeness, or an example of the perfect insect. The appropriateness of the scientific names for the first and third stage does not seem very clear, but there is no doubt of the convenience of having some *one term* by which each different stage of life of any insect may be described; and these are the words that have been adopted; in the following pages some detail is given of these three successive stages of development.

Larva.—Maggot, grub, caterpillar, &c. If an insect-egg about to hatch is held against the light, or examined as a transparent object by means of a strong magnifier, it will be seen that there is a speck inside which increases in size and becomes more regular in shape daily, until it is too large for the egg to contain, when it breaks through this thin film which serves as an egg-shell, and often begins life by eating it. This is the larva. It is usually hatched from an egg, but sometimes is produced alive (as some fly-maggots, or Aphides during the summer months). When it is coloured and has many feet, it is usually called a caterpillar; white fleshy larvæ, such as those of many beetles or flies, are commonly known as grubs or maggots; such as resemble the parent insect are usually known by the name of this insect; but the term of "worm" or "slug" is objectionable, as it leads to confusion.

Larvæ differ very much in appearance: some are legless, cylindrical, or tapering at one end, blunt at the other, with the head (which is soft and furnished with hooks by way of feeding apparatus) capable of being drawn some way back into the maggot; many fly-maggots are of this kind; some larvæ are legless or with a mere

rudiment of a pair of legs on the three rings behind the
head, fleshy, smallest at the tail, and furnished with
distinct head and jaws; such are some kinds of beetle and
wasp-grubs; others are strong and fat, a few inches in
length, with three pairs of legs well developed—as the
Cockchafer grub.

The caterpillars of the butterflies and moths are often
beautifully marked, and have for the most part a pair of
articulated feet on each of the three segments behind the
head, and pairs of fleshy appendages called sucker-feet on
some of the other segments and at the end of the tail,
not exceeding sixteen in all. These "sucker-feet" enable
the caterpillar to hold firmly to the twigs they frequent.
Proceeding onwards still by number of feet, the caterpillars of the Sawflies will be found in almost every case
(Corn Sawfly, *C. pygmæus*, excepted) to have, besides the
three pairs of true feet, five, six, or seven pairs of
sucker-feet, and also the pair at the end of the tail
(known as the caudal proleg). In some cases (as with
Grasshoppers, Aphides or Green-fly, Plant-bugs, &c.)
the young in the first stage—whether produced alive or
hatched from the egg—much resembles the parent, that
is, has a distinct insect shape of head, with horns,
trunk or *thorax*, furnished with six legs, and abdomen;
and differs mainly in size and in being wingless; but,
whether in this shape, or what is known as grub, maggot,
or caterpillar, or whatever kind of insect it may belong
to in this first stage, it is scientifically *a larva*.

In this larval stage the insect feeds voraciously and
often grows fast: the skin does not expand beyond
certain limits, and when this point is arrived at, the
larva ceases feeding for a while; the skin loosens,
cracks, and is cast off by the creature inside, which
comes out in a fresh coat, sometimes like the previous
one, sometimes of a different colour or differently marked.
This operation is known as *moulting*, and occurs from
time to time till the larva has reached its full growth.
The duration of life in the first or larval state is various;
in some instances it only extends over a week or two;

in some (as with the Wireworm and caterpillar of the
Goat Moth) it lasts for a period of three, four, or five
years. As far as observations go at present—that is to
say, with such kinds as have at present been observed—
larvæ are not injured by an amount of cold much beyond
what they are commonly called on to bear in this
country; but they are liable to injury from over supply
of moisture, whether from sudden rain in warm weather
or from full flow of sap of their food-plant, and in this
point of their constitutions we have a principle that may
help much towards getting rid of them. When the
larva has reached its full growth it ceases feeding,
and (in the forms known as caterpillar, grub, or maggot)
it either goes down into the ground and forms a cell in
the earth, or spins a "cocoon" (that is, a web) round
itself of threads drawn from the lower lip (as in the well-
known Silkworm-cocoon), or in some way it makes or
seeks a shelter in which it changes from the state of
larva to that of *pupa*. See figs., pp. 38, 101, 192, &c.

Pupa.—Chrysalis. It is much to be regretted that we
have no generally-adopted word, excepting "chrysalis"
(which is commonly used in the case of butterflies or
moths), to describe the second stage of insect life in which
it is changing from the state of *larva* to that of the
complete insect.

Whilst in this condition it is for the most part without
power of feeding and perfectly inactive, lying (in the
instance of Beetles, Bees and Wasps, and some others)
with the limbs in sheaths folded beneath the breast and
body, or (as with Butterflies and Moths) protected by
a hardened coating secreted from the pores of the
creature within, when it casts its last larval skin.
The method of this change may be easily observed
in the case of the caterpillar of the Peacock-butterfly,
which fastens itself by the tail, and then (after its black
and silver-spotted skin has cracked) by infinite wriggling
and struggling passes this cast-off skin backward, till it
is pressed together at the tip of the tail; and the
creature from within appears in its new form as a bright

green chrysalis, or *pupa*. It is covered with a moist gummy exudation, which quickly hardens and forms a protecting coat, and in due time (if left unharmed) the butterfly inside would crack through this and appear from within the case; but if it is wished to observe that the beginning of the change to the butterfly form has taken place already, one of these chrysalids may be dropped into a little warm turpentine, or turpentine and Canada balsam, directly the caterpillar-skin has been cast; this will soften the gummy coating just mentioned, and the limbs of the future butterfly will be seen. In some cases the change takes place (as with various kinds of flies) in the hardened skin of the maggot, which may be called a "fly-case"; and in some (as with Plant-bugs, Aphides or Green-fly, Grasshoppers, Dragon-flies, and some others) this state of *pupa* is an active one, in which they move and feed, and resemble the perfect insect, excepting in having more or less rudimentary wings or wing-cases. For figures of some of these different kinds of pupæ, the reader is referred to pp. 17, 23, 31, 102, 207, &c.

When the time for development has come, the pupa (if it is one of the active forms, as of a Grasshopper, for instance) may be seen looking heavy and stupid; presently the skin of the back splits lengthwise, and through the opening the perfect insect slowly makes its way out of the pupal skin, carefully drawing one limb after another from its precisely-fitting case, the long hind legs the last, till (in the instance observed, in twenty minutes) the perfect Grasshopper stands by the side of the film of its former self. Flies press out one end of the fly-case, or leave the sheaths of the limbs and body behind (see figure of Onion Fly and Daddy Long-legs). Beetles and Wasps cast the film from their limbs; and Butterflies and Moths crack open the chrysalis-case, and after a short time (during which the wings that had lain undeveloped are expanding) they appear of their full size. The insect is now fully formed; it will grow no more; its internal, as well as external structure is complete; and it is what is known scientifically as the *imago*.

Imago.—Beetle, Butterfly, Wasp, Fly, &c. This is defined as an animal formed of a series of thirteen rings or segments, breathing by means of tubes (tracheæ) which convey the air from pores in the sides throughout the system, and divided into three chief portions. Of these the first is the head, furnished with horns (antennæ), a mouth (differing very much in form in different kinds of insects), large compound eyes (which consist of many small ones formed into a convex mass on each side of the head), and frequently two or three simple eyes on the top.

The second portion (called the *thorax*, or sometimes the "trunk"), is formed of three rings, bearing a pair of legs attached to each, and having usually a pair of wings on the second and third of the rings; but sometimes the wings are wanting, sometimes there is only one pair.

The third portion (called the abdomen) is formed of the remaining nine rings, and contains the organs of reproduction and most of those of digestion.

Insects in this perfect state are of two sexes, male and female; in some instances (as with Wasps and some others) there are imperfectly-developed females, known as "neuters."

After the insect—whether Beetle, Butterfly, Fly, or other kind—has come forth from its chrysalis or fly-case (that is, from the *pupa*), and its limbs have expanded, it grows no more; it is complete, and its remaining work is to support life until it has propagated its species. Usually *pairing* soon takes place, and the male dies; but the female has great tenacity of life until she has laid her eggs. The length of life, however, is various; in some instances a few days, or even hours, is the extent: in others the insects "hybernate," that is, find some shelter in which they pass the winter, and from which they reappear with the return of warmth and sunshine.

CLASSIFICATION OF INSECTS. — Opinions of different writers vary much as to the most desirable form, but the method appears to be the most simple and comprehensive in which they are divided into thirteen orders, arranged according to general similarity in the early stages, and also in the general appearance of the perfect insects composing each order; also according to the number or nature of their wings, or the method in which they are folded beneath the wing-cases.

In the following table the orders are arranged accordingly in the classification given in Prof. Westwood's 'Introduction to Entomology,' these thirteen orders being formed into two great tribes of *Mandibulata* and *Haustellata*, according to whether they feed by means of jaws (mandibles), as in the case of Beetles, &c., or by means of some kind of sucker (haustellum), as is the case with Butterflies, Aphides, &c.

These orders are placed in succession according to the nearest resemblance which the insects of one order bear to the one preceding or following; and the reader will notice that the two last syllables of the name of each order are *Ptera*, meaning "wings," from the Greek word *Pteron*, a wing. The preceding part of the word signifies the nature of the wing.

MANDIBULATA.

COLEOPTERA.—Sheath-winged. Beetles.

EUPLEXOPTERA.—Tightly-folded winged. Earwigs.

ORTHOPTERA. — Straight-winged. Cockroaches, Crickets, Grasshoppers, &c.

THYSANOPTERA.—Fringe-winged. Thrips.

NEUROPTERA. — Nerve-winged. White Ants, May-flies, Dragon-flies, &c.

TRICHOPTERA.—Hairy-winged. Caddice-flies.

HYMENOPTERA. — Membrane-winged. Sawflies, Gall-flies, Ichneumon-flies, Ants, Wasps, Bees, &c.

STREPSIPTERA.—Twisted-winged. Bee-parasites.

HAUSTELLATA.

Lᴇᴘɪᴅᴏᴘᴛᴇʀᴀ.—Scale-winged. Butterflies, Moths.
Hᴏᴍᴏᴘᴛᴇʀᴀ.—Similar-winged. Lanthorn-flies, Cuckoo Spit-
flies, Aphides, Scale-insects, &c.
Hᴇᴛᴇʀᴏᴘᴛᴇʀᴀ.—Dissimilar-winged. Plant-bugs, &c.
Aᴘʜᴀɴɪᴘᴛᴇʀᴀ.—Imperceptible-winged. Fleas.
Dɪᴘᴛᴇʀᴀ.—Two-winged. Gnats, Daddy Long-legs, Gad-flies,
Bot-flies, Flesh-flies, &c.

1. Cᴏʟᴇᴏᴘᴛᴇʀᴀ (Aristotle).—Beetles.

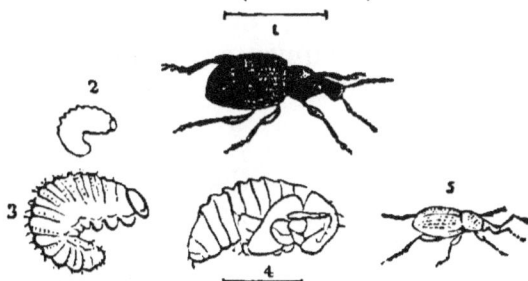

1, *Otiorhynchus sulcatus*; 3, larva: 4, pupa, magnified; 5, *O. picipes;*
fig. 2 and the lines show nat. size.

The upper pair of wings, which are called wing-cases
or *clytra*, are usually horny or leathery, and thus form a
" sheath" for the large membranous under wings which
are folded beneath them. The head is furnished with
large eyes, jaws moving transversely, and with horns
(*antennæ*) of very various shape.

In the water-beetles the hinder legs are often flattened
to a somewhat oar-like shape, and fringed with hairs.

The larvæ are usually fleshy grubs having scaly heads
furnished with jaws: sometimes they are legless, but
commonly have a pair of short legs on each of the three
segments next to the head; and the last segment of the
body (or end of the tail) has often a fleshy foot beneath it.

The pupæ are inactive, of a whitish colour, and
resemble the beetle in shape; with the head bent
forwards, and the legs and wings laid along beneath
the breast and abdomen. For illustrations of various
kinds of beetles, see Cockchafer; Lady-birds; Turnip

c

"Fly," or Flea-beetle; Weevils of various species; and Wireworm Beetle.

2. EUPLEXOPTERA (Westwood).—Earwigs.

The upper wings are *minute* and *leathery*, with the under wings, as the name implies, *tightly folded* beneath. The mouth is furnished with *jaws*, and the end of the tail with pincer-like appendages.

The larvæ and *pupæ* are *active*, and much resemble the perfect insects in shape, except that the larvæ are without wing-cases or wings, and the pupæ, although possessing wing-cases, have only rudimentary wings.

(This order is sometimes known as *Dermaptera*, skin-winged).

3. ORTHOPTERA (Olivier).—Cockroaches, Crickets, Grasshoppers, &c.

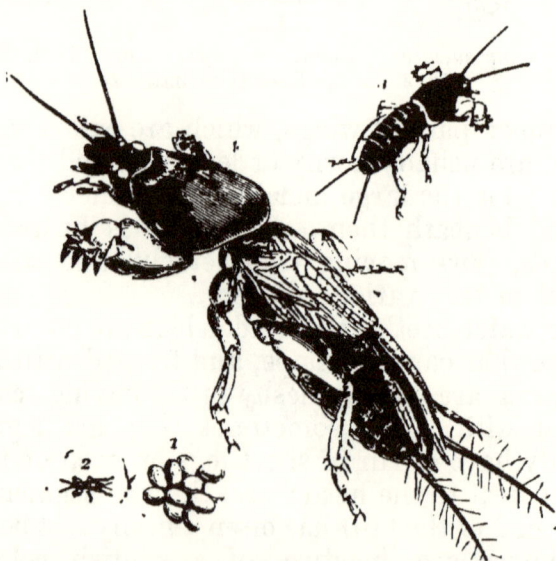

Gryllotalpa vulgaris, Mole Cricket; 1, eggs; 2 and 3, larva, just hatched, and after first moult; 4, perfect insect.

Upper wings *leathery* or *parchment-like*, very *thickly veined*, and *overlapping* at the *tips*. Under wings, which are folded lengthwise or "straight" beneath the upper pair, large, membranous, veined, with larger veins placed somewhat in a fan-shape. The under wings are absent in several species, and the upper wings in one.

Head generally large, upright, with the mouth at the lowest part, and rather backward. Mandibles strong; horns usually thread-like. Legs long and robust. Abdomen joined to the part of the body before it by its whole width, often prolonged at the tail, or furnished in the females with an ovipositor.

The larvæ and *pupæ* are *active* and voracious, and are much like the perfect insect, excepting that the larvæ are wingless and the pupæ have "short rudimental wings and wing-covers which at the first period of this state are but slightly to be perceived."—(J. O. W.)

The Grasshopper is a good example of this order. The Mole-cricket, figured opposite, is sometimes very injurious to plant-life by feeding on the roots, and is remarkable for the size and strength of its fore legs, but its upper wings are not characteristic of the *Orthoptera*.

4. THYSANOPTERA (Haliday).—Thrips.

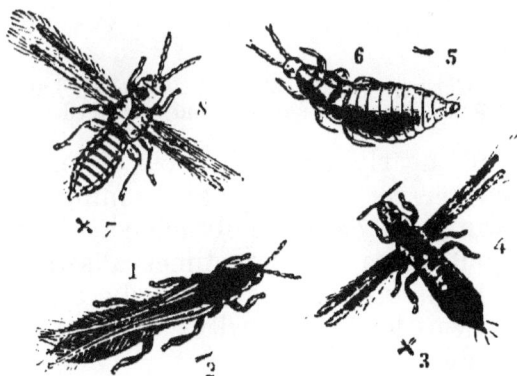

1—4, Corn Thrips (female) at rest, and flying; 5—8, Potato Thrips, larva, and perfect insect flying; all nat. size and magnified.

The insects of this order are often very *minute*, sometimes only about the third of a line in length. They have four wings, which are *nearly alike* and usually long, narrow and with long *"fringes"* all round, laid flat along the back when at rest and somewhat curved outwards. The under side of the head is prolonged into a beak shape, with the parts of the mouth joined into a kind of sucker-like sheath, " out of whose free end the bristle-formed jaws project."

The larvæ and pupæ much resemble the perfect insect in shape, and in the first stage are active, in the second are sluggish.

The Corn and Potato Thrips are examples of this order.

5. NEUROPTERA (Linnæus).—Dragon-flies, May-flies, Stone-flies, &c.

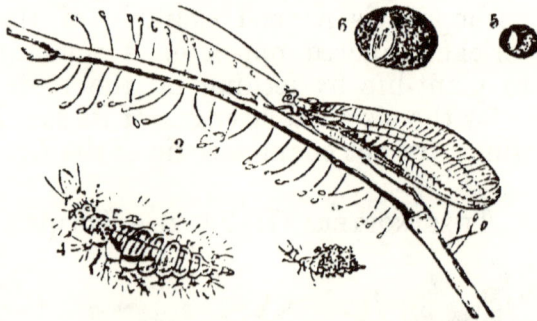

1, Golden-eye, *Chrysopa perla;* 2, egg of *Chrysopa;* 3 and 4, larva; 5 and 6, cocoon, nat. size and magnified.

Wings four, nearly equal in size, membranaceous, with many *"nerves"* sometimes forming a network. The under wings are occasionally folded. Head usually with jaws, but these are sometimes absent, as in the May-flies, which, only living for a short time, do not require apparatus for feeding with. Abdomen generally long and slender.

Larvæ with six legs; pupæ various, in *some cases active*, and somewhat resembling the perfect insect; in others *inactive*, with the limbs folded beneath them.

The families of the Dragon-flies (*Libellulidæ*), Stone-flies (*Perlidæ*), and May-flies (*Ephemeridæ*), pass their first stages in the water, and have *active* pupæ as well as larvæ.

The *Hemerobiidæ* (see *C. perla*, figured opposite) are peculiar in laying eggs fixed by a long stalk of a viscid secretion; the larvæ feed ravenously on Aphides. In the eleven families of which *Neuroptera* is composed, it is said that there is "scarcely a leading characteristic of the order which does not meet with an exception."

6. TRICHOPTERA (Kirby).—Caddice-flies.

Wings four, membranous, the upper usually with branching nerves, and "*hairy*"; the under pair shorter and broader, and folded when at rest; legs long; jaws rudimentary. The early stages are passed in water. The larvæ (known as Cad-baits or Caddice-worms) are nearly cylindrical, with six legs, and live in cases which they form round themselves of little bits of stick, or pebbles, shells, &c.; and in these they change (in the water) to pupæ much resembling the perfect insects.

7. HYMENOPTERA (Linnæus).—Wasps, Bees, Sawflies, Gall-flies, &c.

Humble Bees.—1, *Bombus lucorum*; 4, *B. terrestris*.

Wings four, naked "*membranous*," and furnished with

a few veins; the upper pair (which are much larger than the under ones) are marked on the fore edge with a minute thickened spot called the "*stigma.*" The head is usually furnished with large compound eyes, and also with three simple eyes or "*ocelli*" on the crown; the horns are various (sometimes in the Sawflies with fine comb-like processes). The upper jaws (mandibles) are horny, but not always serviceable for eating with; and in some cases, as with the Honey Bees, a portion of the mouth apparatus (the maxillæ), united with the lower lip and its appendages, form a sucker or proboscis by means of which to draw up their food. The body is covered with a hard skin, and usually has head, thorax, and abdomen distinctly separated.

The abdomen of the female is often supplied with a sting, or with an ovipositor, by means of which she can pierce into animal or vegetable matter to insert her eggs. In some cases this is done by means of a kind of borer; in some, as with the "saw"-flies, by means of a kind of saw-like apparatus.

The larvæ are usually maggot-like and footless, with the mouth commonly but slightly developed; but in the family of the Sawflies (*Tenthredinidæ*, see p. 224) the larvæ are more like Butterfly caterpillars, and have usually, besides the six true feet, twelve to sixteen sucker-feet. (The Corn Sawfly, p. 84, is an exception).

Pupæ are *inactive*, with the limbs of the future insect distinguishable, but wrapped in sheaths and folded beneath the breast. Change sometimes takes place in cocoons.

This order contains the families of Sawflies (*Tenthredinidæ*), Sirices (*Uroceridæ*), Gall-flies (*Cynipidæ*), Ichneumon-flies (*Ichneumonidæ*), Ants (*Formicidæ*), Wasps (*Vespidæ*), and various kinds of Bees (*Andrenidæ* and *Apidæ*), classed under the head *Mellifera.*

For examples, see Humble Bees, Oak Gall-fly, Sawflies of various kinds, and Sirex.

8. STREPSIPTERA (Kirby).—Bee Parasites.

These are named from the small "twisted" appendages which they bear in the place of the fore wings; the true wings are large in proportion to the size of the insect, forming more than a quarter of a circle, with the two straight edges placed in front and against the body. They live in the larval state as parasites in Wasps and Bees, and merely require to be referred to here.

9. LEPIDOPTERA (Linnæus).—Butterflies and Moths.

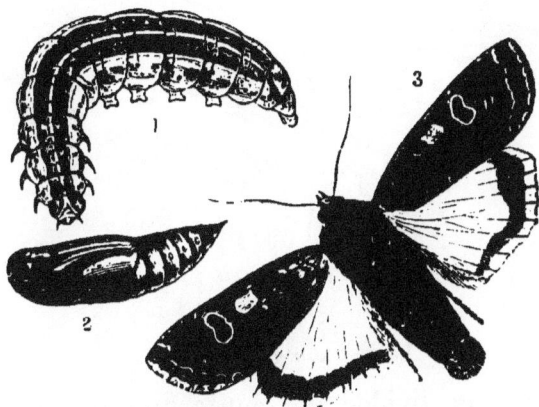

Yellow Underwing, *Tryphæna pronuba.*—1, caterpillar; 2, pupa or "chrysalis"; 3, moth.

Wings four, covered on both sides with fine "*scales*" (whence the name of *Lepidoptera*, from *Lepis*, a scale). Head furnished with two large compound eyes, and sometimes with *ocelli;* horns various; mouth with a proboscis. Body hairy, or scaly.

The larvæ, commonly called caterpillars, are nearly cylindrical, long, soft, and often variously coloured, spined, or tubercled; the head is scaly or horny, bearing eyes, jaws, and a pair of short horns; the three segments next to the head have usually a pair of horny feet on each, and of the remaining nine segments the tail and the *four* intermediate ones are usually each furnished with a pair of sucker-feet or *prolegs*. These vary in number from four to ten, but the pair beneath the tail is seldom missing. The caterpillar when about to change usually spins a cocoon with thread from its mouth, or buries itself, or in some way provides a place of safety; it then moults its skin for the last time, and a viscid moisture exuding from the surface of the newly-exposed chrysalis hardens rapidly over the rudimentary limbs of the future moth or butterfly, and protects it in its *inactive* state, till in due time this outer case is cracked down the back and the insect comes forth.

Large Cabbage Butterfly.

This order is divided into butterflies and moths; the butterflies are distinguishable by their horns being almost invariably thin and long, ending in a knob (see fig.); also by their light and elegant shape, and beautiful colouring. They mostly fly by day, and when at rest carry their wings erect.

Cabbage Moth.

The moths are distinguishable by the horns *never* being club-shaped, but generally thread-like or with side-branches; and they commonly rest with their wings expanded, and are of a heavier make and more sluggish in flight than the butterflies; also, though not exclusively evening or night-fliers, many of them are so.

The family of the Clear-wing Moths (*Ægeriidæ*) is very peculiar, and differs from the other *Lepidoptera* in the wings being more or less transparent or "*clear*," and *without* "*scales*"; but from other characteristics resembling those of this order it has been placed in it.

Hornet Clear-wing.

The larvæ of the clear-wings feed and change to chrysalids in branches or roots of trees.

For examples of Butterflies, see Cabbage Butterflies; of Moths, see those of Cabbage, Turnip, Apple, &c.

10. HOMOPTERA (MacLeay).—Froghoppers, Aphides or Green-fly, Scale-insects, &c.

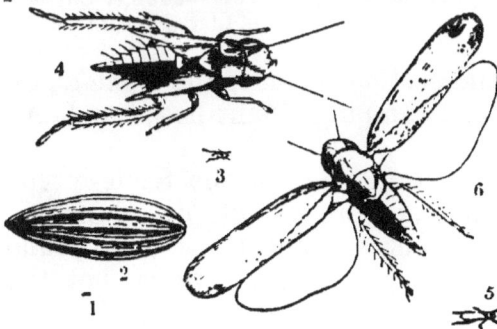

Potato Frog-fly.—1 and 2, eggs; 3 and 4, pupa; 5 and 6, Frog-fly, nat. size and magnified.

Wings usually four (but sometimes two, or absent), entirely membranous, slanting downwards, the upper largest, and not overlapping when at rest. Head with a mouth formed for suction and placed far back beneath it. Horns often short. Larvæ various; pupæ usually active.

This order contains such various forms of insects that it is convenient to take the subdivision, as given by Prof. Westwood, into three sections, of *Trimera*, *Dimera*, and *Monomera*, having respectively three, two, and one, joints in the feet.

The first includes many large foreign insects, as Lanthorn-flies, &c.; and amongst our own the Cuckoo-spit, *Cicada* (or *Tettigonia*) *spumaria*, and the Frog-hoppers, figured on the preceding page.

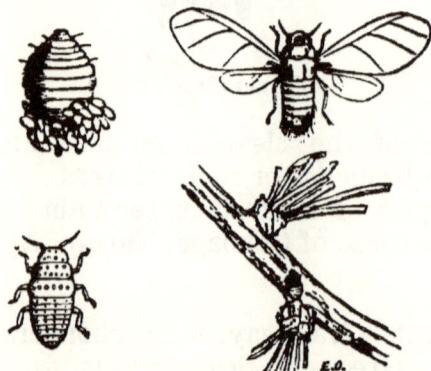

Larch Aphis, *Chermes laricis*, on twig.—Mother *Chermes*, with eggs; winged specimen and larva, magnified.

The second section includes the *Aphides*, or Green-fly; *Psyllidæ*, or Jumping Plant-lice; and *Aleyrodes*, or Snowy-flies.

The third section contains the Scale-insects, amongst which the females are usually fleshy masses, furnished with suckers, but without trace of articulated limbs; he males have one pair of wings, but the mouth is obsolete.

For descriptions and life-histories of the above insects, see references in Index.

11. Heteroptera (Westwood).—Plant-bugs.

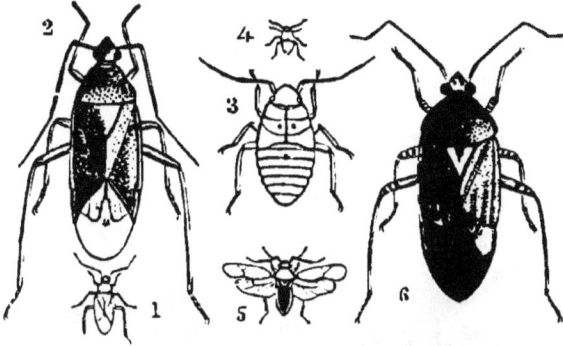

1 and 2, Potato-bug., *Lygus Solani;* 3 and 4. pupa, nat. size and magnified; 5, Hop-bug, nat. size ; 6, ditto, magnified.

Wings four, the upper pair or wing-cases the largest, partly lapping over each other when at rest, and with the part the nearest to the body leathery and *"dissimilar"* in texture to the rest of the wing, which is membranous; under wings membranous.

The head is usually broad, with horns commonly of moderate length, composed of three to five joints; the sucker-mouth is very like that of the *Homoptera*, but placed in front, *not* behind the lower part of the head.

Horns usually somewhat thread-like. Legs various, chiefly formed for walking, but sometimes in the aquatic species with fringes on the hinder pair.

Larvæ resemble the perfect insect, but without rudiments of wings; pupæ with still more resemblance, from these being distinguishable.

One section of this order lives in water, and contains the insects commonly known as Water Scorpions and Water Boatmen; the other, besides the "Water-measurers" common *on* water, contains various kinds of Bugs preying on plant and animal life by means of their suckers, and characterized generally by a power of giving out a scent on being alarmed, which is usually, but not always, of a disagreeable kind.

12. APHANIPTERA (Kirby).—Fleas.

Four scales, which are *"imperceptible"* to the naked eye, take the place of wings. The legs are long and formed for leaping, and the mouth for suction.

The larvæ are minute worm-like, footless grubs; the pupæ are inactive, with legs enclosed in sheaths.

13. DIPTERA (Aristotle).—Flies.

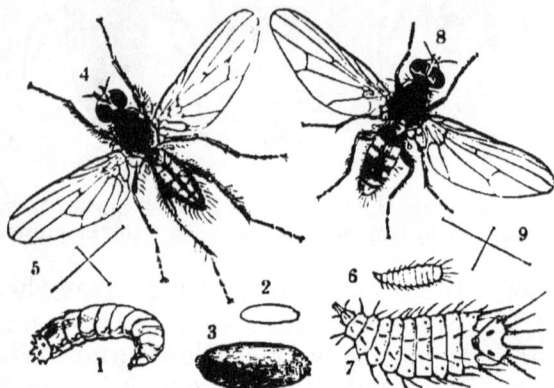

Various kinds of dipterous flies (see pp. 31—34).—1, 6, and 7, larvæ; 2 and 3, pupæ, nat. size and magnified; 4, 5, 8, and 9, flies, magnified, with lines showing nat. size.

Wings *"two,"* membranous; in the place usually occupied by the hind wing are a pair of slender filaments with a knob at the end, called "poisers" or *"halteres."* Head usually distinct, and horns generally inserted near together on the forehead. Mouth formed for suction. Legs long.

Larvæ fleshy, cylindrical, and footless, but sometimes with indications of feet; and it is only in this order that the head of the larva is sometimes soft and fleshy; mouth generally furnished with two hooks, as an apparatus for feeding with.

Pupæ various; in most cases the skin of the larva shrinks and hardens, so as to form a case in which the change takes place, and out of which fly-case the fly makes its way when developed by cracking off one end;

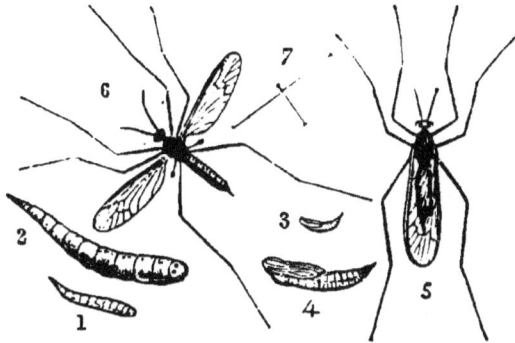

Winter Gnat, *Trichocera hiemalis;* larva and pupa, nat. size and magnified.

in the case of Gnats (*Culicidæ*) the pupæ are active, and swim in the water; the pupæ of the Daddy Long-legs and of the Winter Gnat, figured above, which are of the family *Tipulidæ,* show in some degree the form of the contained insect, and have the limbs enclosed in sheaths; and in the case of the "Forest-flies" (*Hippoboscidæ*) the insect passes the larval stage and changes to the pupa in the abdomen of the female before being deposited.

This order includes Gnats, Daddy Long-legs, Gad-flies, House-flies and Flesh-flies, Bot-flies, &c.

For illustrations of Cabbage and Carrot Flies, Daddy Long-legs, Onion Fly, Turnip Leaf-miner, &c., see Index.

PART I.

FOOD CROPS

AND

THE INSECTS THAT INJURE THEM.

FOOD CROPS

THE INSECTS THAT INJURE THEM.

---◆---

ASPARAGUS.

Asparagus Beetle. *Crioceris asparagi*, Linn.

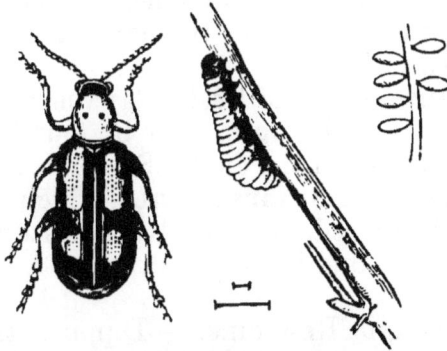

Asparagus Beetle, larva and eggs; all magnified. Natural length of egg
and beetle shown by lines.

The ASPARAGUS BEETLE often causes injury, and in
some seasons does much damage, by the grub eating off
the leaves of the Asparagus, and gnawing the more
tender shoots so as to destroy them.

The *eggs* are dark-coloured, somewhat spindle-shaped,
and may be readily observed fastened by one end along
the shoots, or on the unopened flower-buds.

B

The *grubs* are of a dirty olive, or slate-colour, and exude a large drop of blackish fluid from the mouth on being touched. From the tail being curved and holding fast by a fleshy foot, it is very difficult to pick them off. They are full-fed in a fortnight, when they go down into the ground, spin parchment-like cocoons, in which they change, and come up as perfect beetles in about another fortnight or three weeks.

The *Beetles* are about a quarter of an inch long, blue-black or greenish; the body behind the head red, with two black spots. The wing-cases are ochreous-yellow, with a line down the centre of the back, a branch from each side of it, and also a spot or patch at the base and tip of each wing-case of blue-black. These markings form a kind of cross, whence the name sometimes given of "Cross-bearer."

The successive broods of Beetles lay their eggs directly, and the insect, in all stages, may be found from about the middle of June till the end of September.

The 12-spotted Asparagus Beetle, which differs from the above in being red, with twelve spots on the wing-cases, is seldom found in England.

Prevention and Remedies. — Dipping the infested shoots in a mixture of half a pound of soft-soap, a quarter of a pound of flower of sulphur, and about the same quantity of soot, well mixed together in a pail of warm water, has been found a good remedy. The infested shoots were well dipped, and next day the grubs were found to have all been cleared off. The plants were afterwards syringed, to clean off the dirt left by the dipping, and soon regained their healthy appearance. —(W. D. C.)

Syringing with water warm enough to make the grubs loosen hold, but yet not of a heat to hurt the leafage, clears them off well. The greater part of the grubs fall as the water touches them, and the rest on a smart tap being given to the shoot. Dry soot plentifully thrown on to them, whilst they are on the ground under the

Asparagus and are still wet, appears quite to prevent any return of the grubs to the shoot. If any, either of the beetles or grubs, return, a second treatment would probably clear them completely.

A large number of beds may soon be dressed by a man and boy going round together—one syringing, the other striking the shoots and throwing the soot upon the grubs; and the growth of the Asparagus, after this slight manuring, is good. The water should not be of a heat above what can be well borne by the hand.

Hand-picking has been recommended, but from the strong hold that the grub has on the shoot this is difficult to manage without hurting the plant; but a little salt (or any other application disagreeable to the grub) taken in the fingers, instead of working bare-handed, helps to make it loosen its hold.

Cutting off the shoots that are badly infested with eggs and burning them is of service.

Shaking the beetles into a wide bason or tray, held below to receive them, has been recommended; but it is desirable that some mixture of mud or soot and water (or of a nature that may stick to them at once) should be provided for them to fall into, or they would soon escape.

Strewing the plants well with unslaked lime early in the morning, whilst the dew is still on, is also recommended.

BEAN.

Bean Aphis. { *Aphis rumicis*, Fabr.
         ,,   *fabæ*, Kirby & Spence.

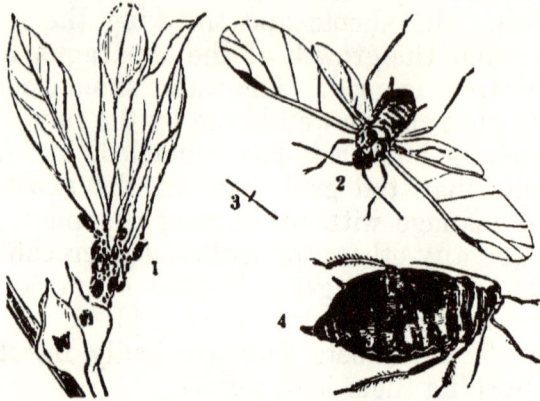

Bean Aphis: 1, Bean-shoot. with Aphides; 2, male, mag.; 3, nat. size.; 4, wingless female, magnified.

The BEAN APHIS (known also as "Black Fly," "Collier," and "Black Dolphin") sometimes appears in such vast numbers as to smother the Beans, making them look as if they were coated with soot. The attacks are begun by a few wingless females establishing themselves near the top of the Bean-shoots, where they produce living young. These in their turn are soon able to produce another living generation; and so on, and on, till the increase is enormous, and from the numbers of the "Black Fly," and the sticky juices flowing from the punctures which they have made with their suckers, the plant becomes a mere dirty infested mass, with a few diseased leaves sticking out from amongst the plant-lice.

These Aphides are very similar in shape throughout all their stages, excepting that in the first and second (which answer to those of larva and pupa), although they have six legs and are active, they have not wings.

The young are slaty gray, but soon acquire a blackish velvety coat. The pupa is larger and slaty gray, excepting the abdomen, which is black with various white spots, and the cases for the wings are also black.

Of the two kinds of females producing living young, the wingless kind is shiny (or sometimes dull) black, excepting the shanks and the middle joints of the horns, which are somewhat ochreous. The head, body, and abdomen are so much grown together as to seem almost like one piece (see fig. 4, magnified).

The winged female is shiny and black, or with a brownish tinge; the shanks and middle joints of the horns are amber-yellow, and the wings are yellow at the base, with a green line and mark on the fore edge, and brown veins.

The egg-laying female—that is, a third description, which appears in autumn, and lays the eggs from which a new series of generations, producing living young, will start again in the following year—is very like the wingless viviparous form figured above. The male (fig. 2, magnified) is black and winged.—('Farm Insects,' and 'Mon. of Brit. Aphides ').

PREVENTION AND REMEDIES. — No better plan appears to be known than cutting off the infested tops of the Beans, but it should be done as soon as the "Colliers" are noticed. This point is very important, on account of the extraordinary rate at which Aphides increase. The tops should be trampled on thoroughly as they are cut off; sheared into baskets and burnt; or destroyed in some other way with the Aphides on them, for if these are left amongst the Beans the insects are able to make their way back again to the growing plants, and thus little good will have been done.

In garden cultivation, where there is only a small amount of crop to be attended to, it sometimes answers to throw soot on the infested plants. This lodges well amongst the Aphides and in the axils of the leaves. Any dry dressing that would thus lodge, and make the

Bean-tops more or less unpalatable to the Aphides, would be of use. Where only a small extent of crop needs attention, a good drenching with strong soap-suds, or syringing with a solution of soft-soap, would also be useful ; the soapy matter sticks to the Aphis, and is thus a much surer remedy than many of the attempted applications which run off at once from the skin of the insects, and consequently are useless.

A healthy, luxuriant growth is also of importance. The "Colliers" may attack the healthy as well as the unhealthy plants, but the strong growth which is run on by previous good cultivation of the ground, and also the application of a little liquid manure if desirable, will keep the plant in heart with a plentiful flow of sap, and thus it will suffer much less from attack than the weakly, stunted growths that have no power to replace the juices which the Aphides are constantly sucking out of them.

Removal of the wild plants on which this kind of Aphis is to be found would probably materially lessen its numbers. These "Colliers" or "Black Flies" are to be found in great numbers on the Curled Dock (*Rumex crispus*, whence their name of "*rumicis*"), and also on Thistles ; and it is stated that the wingless egg-producing female winters in Furze-bushes, the branches of which may be found dotted with Aphis eggs.—('Farm Insects,' 'Brit. Aphides,' &c.)

| Humble Bees. | Earth Bee, | *Bombus terrestris.* |
| | Wood Bee, | ,, *lucorum.* |

These Humble Bees sometimes do harm by piercing a hole through the lower part of the calyx into the Bean-flower, so as to get at the honey inside ; and by so doing they cause a part, or even the whole, of the Bean-pod to prove abortive. They have been known to do serious injury ; but it, generally speaking, may be doubted whether the larger size of the pods, from a few of the blossoms being thinned out, does not make up for the slight damage.

The female, or Queen of the Earth Bee, is black, with a band of orange-yellow just behind the head and another on the second ring of the abdomen, and with two tawny bands near the tip of the tail. The worker has the yellow often of a paler colour, and the bands toward the tip of the tail white instead of tawny. The male or drone has the orange-yellow brighter than the female, and the tip of the abdomen pale tawny or white.

1, Wood Humble Bee; 2 and 3, punctures made by Bees in Bean-calyx; 4, Earth Humble Bee, sucking nectar.

The Queen of the Wood Bee is smaller than the above-mentioned queen, and is black, with a band of lemon-yellow just behind the head and on the second ring of the abdomen; the three end segments of the abdomen are snow-white. The worker is marked like the queen. The drone has a black band between the wings and across the abdomen, and has (like the queen and worker) the three rings at the tail snow-white; otherwise it is for the most part yellow or yellowish white.

The nests may be found in old loose walls, heaps of rough stones, in dry banks, at the roots of trees, and such-like places. The nests of the Earth Bee may be found occasionally at the distance of five feet from the entrance of the hole.—(Descriptions from 'Bees of Great Britain').

PREVENTION AND REMEDIES. — If any serious injury is found to occur, the best plan is to have the nests searched for and destroyed. For a few pence any boys that are about will soon find them out ; but unless it is quite certain the Bees are doing harm, it is much better not to destroy them.

Bean Beetle. *Bruchus granarius,* Linn.

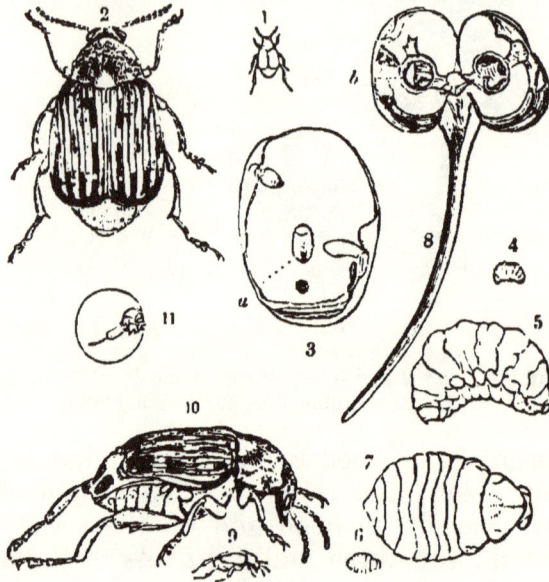

1 and 2, *Bruchus granarius*, nat. size and mag.; 3, infested Bean split open, showing cell ; 4 and 5, larvæ, nat. size and mag. ; 6 and 7, pupæ, nat. size and mag. ; 8, Bean injured by Beetle, vegetating ; 9 and 10, *Bruchus pisi*, nat. size and mag.; 11, injured Pea.

The BEAN BEETLE injures the crop by laying its eggs in the Beans and Peas whilst they are still soft in the pods, often choosing the finest for the purpose.

The maggots feed inside the seed, sometimes eating away most of the contents, but generally leaving the growing germ uninjured so that the seed does not lose its power of sprouting.

When full-grown, the maggot gnaws a round hole to the inside of the husk of the seed, and usually cuts a line round this kind of circular lid of its burrow, so that when needed to be displaced afterwards, for the escape of the Beetle, the bit will fall out on a touch. The maggot turns to the pupa in the Bean or Pea in which it has fed, and appears generally to pass the winter in this state. When spring comes it changes to the perfect Beetle, and makes its escape either by gnawing a way out, or—commonly—by pushing out the loose lid of its burrow.

The Beetle is little more than the eighth of an inch long, and if looked at under a magnifying glass will be found to have the head drooping, with the mouth forming a kind of wedge-shaped beak, the fore part of the body somewhat bell-shaped, and each wing-case pitted with ten rows of small dots. The colour is black, with brown hairs and various white spots; the tip of the tail prolonged, and covered with grey down. The front pair of legs are reddish. The Beetles may be found on flowers of the Furze as early as February, but though they may be found in seed beans in March, April, and May, they do not always leave the seed as soon as they are developed.—(J. Curtis.)

PREVENTION AND REMEDIES.—A great deal may be done in the way of prevention by attention to the seed before it is sown. The maggot goes on feeding in the seed after it is stored; by the end of autumn probably all the beetle-maggots will be full-fed, and will have eaten their way in the seed to the inside of the skin. This will show on the outside as a round or oval mark (about the size figured at 6 *a*), rather duller in colour than the rest of the skin, and rather more transparent, from the substance of the seed being cleared away behind it. If this round piece of skin be lifted off, the Bean Beetle will probably be found inside, and by this mark infested seed may be known. Such seed should not be sown. If Beans or Peas thus infested are sown, the Beetles will not be in the slightest degree injured by being buried, but will come up through the ground in due time to infest the new crop.

The Bean Beetles begin to appear in February, although some may still be found in the seeds till May. When they have left the seed, the round hole through which they escaped will show where they have been, and such seed is to be avoided. The injured seed will sprout in most cases, but although the growing germ is left, a great part of what this germ needs to make it grow healthily is gone. If we consider that, in germination, chemical changes take place by which the insoluble starch in these seeds is turned into soluble plant-food, and that on this alone the young sprouting plant is nourished until the leaves and rootlets are produced, it will be evident that the chances of a strong and healthy plant are much diminished by using maggot-eaten seed. The young plant depends on the quantity of food in the seed for the vigour of its first start, and if this first start is not vigorous the later growth will probably be stunted and sickly also.

Crops of autumn-sown Beans have been found to be the most infested, because, as above mentioned, the Beetles are still in the seed, and before the maggot-injury shows through the skin it is not easy to tell which are the infested Beans. It is difficult to apply any remedy in the field suitable to such a small insect, but the use of spent Hops as a manure, which is found serviceable in other cases of insect-attack, might be of use here.

Bean Beetles (*Bruchi*) of various species are found— sometimes in enormous quantities—in imported seed, and where there is much Bean-growing round mills where Beans are ground, it would be highly desirable some measures should be taken to save us from the consequences of the vast numbers sometimes to be found in such places.—(See 'Farm Insects,' note, p. 363.)

Dipping the Beans or Peas in boiling water for one minute is stated to kill the grub inside without hurting the seed, but as dipping for four minutes generally destroyed the germinating power the experiment is much too hazardous for general use as regards seed, though it might be acted on with advantage with regard to Beans or Peas about to be ground.

The simple remedy practised in the Colonies for the Weevil in Rice,* which destroys life by exhausting the air, might be of service in this case; and also it is probable that if greater care was exercised in choice of seed by the buyers, especially in the retired spots and small holdings where "cheap seed" finds a market, that this "pest," which is not so troublesome as many others, would be materially lessened.—('Farm Insects,' &c.)

* I procured a large tank, such as is kept on ship-board for the water supply. In it the Rice was put, filling it up until a short distance from the top. Upon the top of the Rice is set a candle or lamp alight, after which the opening is closed and rendered air-tight by the use of a small quantity of white-lead. Next day the tank may be opened, when the whole surface will be found to be covered with dead Weevils.—G. B.

BEET.

Beet Carrion Beetle. *Silpha opaca*, Linn.

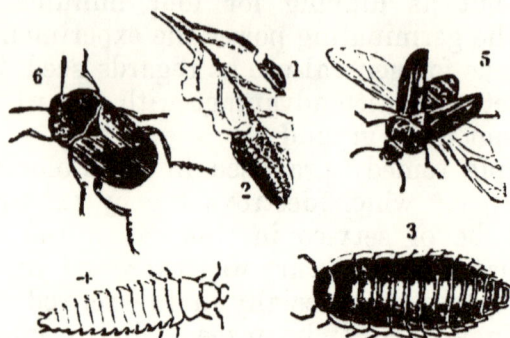

1 and 2, Young and full-grown larvæ; 3 and 4, larvæ magnified;
5, female Beetle flying; 6, male Beetle, slightly magnified.

This BEETLE is often to be found in dead animals, and
its grub was supposed to live only in putrid matter (as
in dead birds, rabbits, hedgehogs, and the like), till in or
about the year 1844 it was found feeding on leaves of the
Red Beet in France, and the grub of a *Silpha*—apparently
of the same species and certainly similar in its habits—
was observed in the same year feeding in the Mangold-
Wurzel crops in Londonderry. At this date, and in
several of the years following, they ruined whole crops
by attacking the leaves, sometimes as soon as they
appeared above ground; in other cases they did harm
by checking the growth of the root till the leaves sprouted
again. The leaves were gnawed away until only the
fibres remained, but the roots do not appear to have
been meddled with.

The grubs are much like Wood-Lice in shape, between
a third and half an inch long when full-grown; the three
rings or segments next to the head are rounded at the
sides, but in the other segments they are sharp, so as to
give the sides of the grub a saw-like appearance, and the
tail segment has a sharp spine on each side. These

grubs are black and shiny, sometimes with a little yellow at the front edge of the segment. When full-fed (which in the instances noted was about the end of June) the grubs bury themselves and form cells at the depth of three or four inches below the surface of the earth, in which they turn to pupæ, and from these the Beetle has been seen to come up in about the space of a fortnight or three weeks.

These Beet "Carrion" Beetles, as they are called, from the mixed nature of their food, are flattish, of the shape figured magnified at "5" and "6," about five lines long, brown-black, with a tawny down which easily rubs off, when the Beetle appears as black (except the tip of the abdomen, which is dull red) and pitted all over. The eyes are large and oval, the horns club-shaped. The body behind the head is twice its width, and somewhat oval. The wing-cases are very flat, and turned up at the outer edge. Each wing-case has three sharp ridges running along it, the middle and outer ridge having a raised lump between them. The tip of the abdomen is dull red.

These Beetles have large wings folded under the wing-cases, and besides being common in April, in the putrid matter that they like, they are to be found at the roots of trees, under stones, and in the flowers of the Mountain Ash.—('Farm Insects.')

PREVENTION AND REMEDIES. — More information is needed as to the life-history of this Beetle before it is possible to tell how to counteract it.

The egg in common circumstances is laid in putrid matter, and in the case of the attacks occurring to the Irish crops in the neighbouring counties of Londonderry and Tyrone, it appears not unlikely that some kind of manure particularly attractive to the Beetle was in use in those years in the district, although Curtis's enquiries only elicited that in the case where the attack accompanied the use of farm-yard manure, that "no more bones and offal, beyond what might accidentally be thrown on a dunghill, had been used." It is worth observing that where some Beet-roots had accidentally been left in ploughing up the injured crop, these shot leaves again and recovered. This points, first, to the prudence of not giving up hope too soon, and unnecessarily destroying a crop; and also to the great importance in attacks of this kind, which only last for a few weeks, of keeping up the strength of the plant by any available means, whether by the application of manure, irrigation, or otherwise, till in the natural course of their lives the power of the insect to do harm has gone by.

In the case of this grub the attacks are given as occurring from the 21st of May to the 24th of June, and in another case from a little after the beginning till the end of June; and no mention is made of any damage from the summer brood of Beetles.

Looking at such information as is given, it appears probable that if farm-yard manure was applied to the soil intended for Beet in the autumn instead of the spring, it would be a good means of prevention. The after cultivation would so dispose of the putrefying matter that there would be little to attract the Beetle to lay its eggs. Manure used near sowing-time should for the above reason be well rotted and free from all putrefying offal. The substitution of artificial manure would be a good preventive, where attacks often occur.

Should attack take place, dressing the plants with superphosphate would be likely to do good ; or a good sprinkling of the mixture of lime, gas-lime, soot, and sulphur—of which a note is given under "Turnip Fly"— would probably both check the insects and encourage the growth of the crop.

Beet or Mangold Fly. *Anthomyia betæ*, Curtis.

The BEET FLY damages the crops by means of its maggots, which feed on the pulp of the Beet or Mangold leaves, and thus reduce the leaves, or large patches of them, to nothing but dry skin. This kind of attack had been little noticed in this country till last year (1880), when the "Mangold maggot" was prevalent in many localities, and especially in Westmoreland and Cumberland, where, of 1624 acres of Mangolds grown in those counties, it is reported that all were infested.—(H. I. L.)

The eggs are very small, snow-white, and oval (see fig., much magnified, by Farsky, showing honeycomb-like markings). These are laid in small patches beneath the leaves ; as soon as the maggots are hatched they bore through the skin of the leaf, and, being voracious, clear away the substance rapidly. The maggots are about the third of an inch long, legless, cylindrical, blunt at the tail, and tapering to the head, which carries two

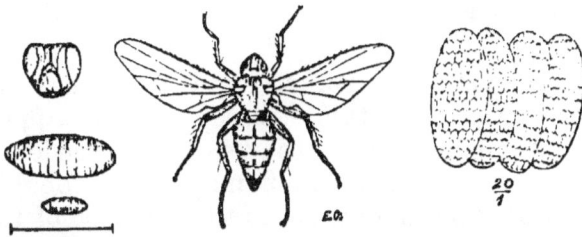

Beet Fly (female), mag. ; line showing spread of wings, nat. size ; head, mag. Pupa, nat. size and magnified. Eggs (after Farsky), mag.

black hooks by means of which it cuts away the pulp of the leaf. The colour is yellowish white, but sometimes green, especially towards the tail, from the intestines filled with green food showing through the thin skin. They feed for about a month, and then turn to chestnut-brown pupæ, sometimes in the leaves, but commonly they quit the leaves and pupate in the ground about three inches below the surface and near the attacked plant.

In summer the flies come out in about ten days or a fortnight. These are ashy grey, with various darker markings, and with black bristly hairs ; and about half an inch across in the spread of the wings. In the male the large brown eyes nearly meet on the top of the head, and the abdomen is very narrow, straight along each side, and with a black triangular spot down the centre, at the base of four of the segments. The female (figured above, magnified) has the eyes distant, with a narrow white band round them on the face, and the abdomen is enlarged at the base, tapering to a blunt point at the tail, with three indistinct dark stripes along it.

The Flies appear from March to May, and there are two broods (if circumstances are favourable there are probably more) in the summer and autumn. The date of development of the latest brood is variable ; in regular course it passes the winter in the pupal state, but sometimes specimens develop at once, and hybernate.— ('Notes of Inj. Insect Observation,' 1880, and 'Entomologist,' Jan., 1881.)

PREVENTION AND REMEDIES.—As this Fly has not been generally noticed in this country till its appearance in 1880, remedies seem to be little needed. The unusually wet season of 1879 may very probably have been exceptionally favourable to the increase of this Fly as it was to that of the Daddy Long-legs (*Tipulæ*), another of the same class of two-winged Flies turning to pupæ in the ground.

It is noticeable, however, that especially bad attack is reported from peat-land ; and also from "fen-like soil" near the shore of the Solway Firth.—('Journ. R. Ag. Soc.')

In a case of attack reported from Cheshire, different parts of the field were dressed experimentally with guano, with soot, and with mineral superphosphate ; and all had a good effect, but the superphosphate was the best.—(S. F.)

The Silver Y Moth. *Plusia gamma*, Linn.

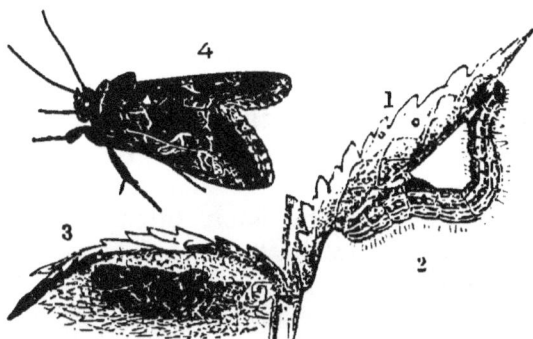

Moth, caterpillar, and pupa in cocoon.

This moth seldom causes much injury in England, but as it is always more or less about during the summer and autumn, and sometimes appears in vast numbers, it is desirable to notice it.

It is widespread in its localities; it is to be found over the northern half of the globe from Abyssinia to Greenland; it is said to extend to the frontiers of China and Siberia, and it is also prevalent in North America.

In 1735 the caterpillars did much damage to Peas and Beans in market gardens round Paris; in 1816 the moths were noticed in vast numbers in the northern part of France. In 1831 the caterpillars appeared in Bavaria, and in 1868 on the Sugar Beet in Saxony; and again in 1879 they appeared in Saxony in such overwhelming numbers that it is shown (from statements quoted in the 'Times' of Nov. 12th, 1879, p. 6, col. 1, of information given at the previous meeting of the Society for the Promotion of Sugar Beet Industry at Halle) that— "Before the appearance of the moth and caterpillar, the Sugar Beet crops in Saxony were in excellent condition, and would in ordinary circumstances have yielded a harvest of from nine to ten tons per acre; the actual

c

yield, where the caterpillars had been, was only three tons."

The Moths generally lay their eggs beneath the leaves " in considerable clusters" (J. C.), singly, and hatching in ten to fourteen days (E. L. T.); and although the caterpillars may be seen feeding by day, they are not easily to be observed, from their colour resembling the leaves. When full-grown they are green, with a green or brownish head, narrow white streaks along the back, and a yellow streak along each side; and are covered with short hairs. They have a pair of reddish brown feet on each of the three segments behind the head, but have only two pairs of sucker feet beneath the body, and one pair at the tail, these are all green. When full-fed they spin a woolly or silken cocoon in a leaf or on a stem, within which they change to a pitch-coloured chrysalis. The moths have the head, body between the wings, and crests running along the back and abdomen, of a purplish brown or deep ashy grey; the fore wings, which have a satiny lustre, are variously tinted with grey and brown, the distinguishing mark being a bright white or yellow figure resembling the letter of the Greek alphabet known as "Gamma" and the English "Y," whence the name of "Gamma" or Y Moth. The hinder wings are whitish, with dark veins, and a broad brown margin. The Moths may be seen as early as April, but are most common towards autumn. —('Farm Insects,' 'Brit. Moths,' 'Prak. Insecten-Kunde,' &c.)

PREVENTION AND REMEDIES. — These caterpillars feed on most of the low-growing plants, and also, if pressed for food, on the grasses; so that clearing weedy or grass-grown spots in or round gardens is a good means of prevention; nettles especially should be removed, as these are one of their food-plants.

When the caterpillars are seen, a dusting of caustic lime, soot, or salt, is a good remedy; and hand-picking a sure, but very troublesome one.—(M. D.)

Shaking the infested plants or leaves (as may easily be done by a light blow with a small bough or birch-besom) so as to make the caterpillars fall off, and then trampling on them, is a good remedy.

Drenchings of liquid manure, or of water alone thrown over the plants would be serviceable, from the circumstance of sudden wet being prejudicial to the caterpillars; and also—as from their great size they consume the leaves very rapidly—it is very important to stimulate the growth of the attacked plants as much as possible. Should a serious attack occur to field crops, many of the methods of remedy given in detail for caterpillars of Moths, or of Sawfly on Turnips, would be applicable.

CABBAGE.

Cabbage Aphis. ⎫ *Aphis brassicæ*, Linn.
   ,,   Green Fly. ⎭

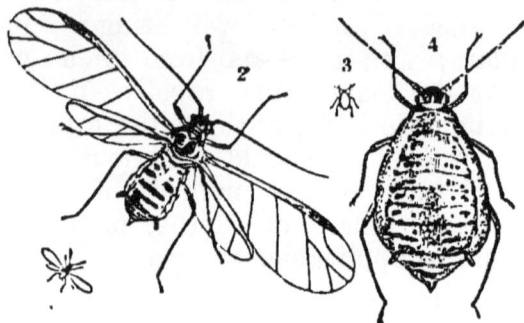

1 and 2, Male Aphis; 3 and 4, female (nat. size, and magnified).

CABBAGE APHIS.—These "Green Flies" may be found on the Cabbage in great numbers during the summer, clustered under the outer leaves, and also in the folds or on the upper side of the inner leaves; and some may be found remaining even as late as the end of November.

They do much harm by inserting their suckers in the plants and drawing away the juices; and also causing a much deformed and diseased growth.

The young, when first born, are yellow. In the next stage (which answers to the *pupa* one, and whilst as yet they are without wings) they are much wrinkled, of a dirty green colour, with olive-green or grey-black wing-cases. The wingless females are mealy, and when this meal is rubbed off they are of a greyish green, with black spots on each side of the back; eyes and legs black; horns green or ochreous, with black tips.

The winged female is of a yellower green, with head and markings between the wings black, and some dark marks across the abdomen. The legs and horns dark brown, and a mark on the fore edge of the wings also dark. The male is given by Curtis as pea-green, otherwise it is much like the winged female, excepting in the mark on the wings being green.—(' Farm Insects,' and ' Mon. of Brit. Aphides.')

PREVENTION AND REMEDIES. — In garden cultivation, drenching the infested plants with soap-suds is practicable and of service, especially in killing the young Aphides.

Syringing with an infusion of tobacco mixed with lime-water has been found very useful, and the following mixture is also stated to be serviceable :—Four ounces of quassia boiled for ten minutes in a gallon of water, and a piece of soft soap about four ounces in weight then added ; and the mixture syringed over the plants. In this application the soft soap is the important matter. From the mealy or powdery nature of the coats of the Aphides, mere waterings are apt to run off from them harmlessly, and adhesive applications like syringings of soft soap are much surer remedies.

Thorough drenchings of water with the garden-engine, however, are of service, by forcibly clearing many of the Aphides from the plants and also by encouraging growth.

We all know the overwhelming increase of Green Fly that often happens when the plants are stunted by heat or disease,—or by the Aphis attack itself ; and it is noted (see ' Mon. of Brit. Aphides,' by G. B. Buckton, F.R.S., vol. i., p. 72) that when the juices of the infested plants begin to fail and become sickly from excessive numbers of Aphides, a change commences in larvæ subsequently born. Signs of wings appear, and the viviparous females from these pupæ are winged; this different development, with its increased power of spreading attack, following apparently on the altered food.

This is curious as a scientific observation, and, if always the case, watering would do much good by keeping up the flow of sap; and applications of liquid manure and such methods of cultivation generally as will keep the plants in vigorous health are to be advised, both as making the plants less suitable for the insects and also preserving them from important injury by any excepting severe attack.

A careful dusting with caustic lime or soot is also very effective in getting rid of the Aphis, and some amount of good may be done by carefully breaking off the leaves that are coated with Green Fly (as happens in a bad attack) and crushing them under foot, or putting them as they are gathered into a sack, so that they can be thrown either under water in the sack, or out into wet manure. Any way that will kill them at once will do. In the case of Aphides (as also with other insects of which one kind infests many kinds of crops), the remedies are mainly given in connection with the crops that are most attacked; the reader is referred, for further details of Aphides and remedies, to Hops, Turnip, Plum, &c.

Large White Cabbage Butterfly. $\left\{ \begin{array}{l} \textit{Pontia brassicæ,} \text{ Curtis.} \\ \textit{Pieris brassicæ,} \text{ Latr.} \end{array} \right.$

The caterpillars of this Butterfly are very destructive to the Cabbage crop by eating away the leaves until at times nothing is left but the large veins; they do serious damage to White Mustard by feeding on the pods left for seed, and, in the case of Turnips, they feed on both the leaves and seed-pods.

The eggs are bright yellow, and are laid in clusters beneath the leaf.

The caterpillars are greenish at first, afterwards bluish or greenish above, yellow below, with a yellow line along the back, and another showing more or less plainly along each side; they are spotted with black, and have tufts or a sprinkling of hairs. When full-fed they wander off to some protected place, under old boards, beneath eaves, in open sheds, or the like places, and there hang themselves up by their tails and turn into chrysalids of a pale green colour spotted with black. The butterfly comes out in about a fortnight from the midsummer brood of chrysalids, but not till the following spring from the chrysalids that form in autumn. The fore wings of the

"Large White" butterfly are white on the upper side, with a broad black patch at the tip, more or less notched on the inner side; in the female there are also two black spots in the middle of the wing, and a blackish splash at the hinder margin; the under side is white, with a yellow tip, and with two black spots, both in the male and female. The hind wings are white above, with a small black patch on the front margin, and beneath they are of a dull palish yellow, speckled with black.

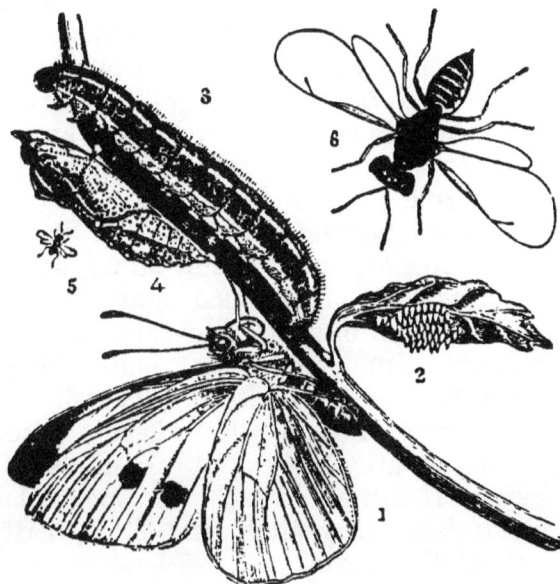

1, Female butterfly; 2, eggs; 3, caterpillar; 4. chrysalis; 5 and 6. parasite Ichneumon-fly, *Pteromalus brassicæ*, nat. size and magnified.

PREVENTION AND REMEDIES.—The habits of this species of Cabbage Butterfly and of the "Small White" are so much alike that the following methods of treatment are applicable in either case, excepting with regard to destruction of the eggs :—

The "Large White" lays its eggs in clusters beneath the leaves, and as soon as the Butterflies are noticeable, the eggs should be looked for, and the pieces of leaf covered with them torn off and destroyed. The eggs of the "Small

White" and "Green-veined White" are laid singly, therefore this method of treatment is not applicable.

A most serviceable way to lessen the numbers of this pest—and one easily managed—is to have the chrysalids searched for and destroyed.

It has been observed that the proportion of injury from attack of Cabbage caterpillars is much larger to Cabbage crops grown in gardens where there are plenty of protected places, such as the caterpillar chooses for its change to the chrysalis state, than to the crops grown in fields where such shelter is not at hand.

When the first brood of caterpillars are full-grown, and have disappeared from the Cabbages in early summer, they have left them to turn to chrysalids in any sheltered nook near, and may be collected in large numbers by children for a trifle per hundred. They may chiefly be found in outhouses, potting-sheds, and the like places, in every neglected corner, under rough stairs, step-ladders, or beams, or shelves; or fastened against rough stone walls or mortar. Out of doors they may be found under eaves, or palings, or under pieces of rough timber, broken boards, or any kind of dry sheltering rubbish.

It is very desirable not to allow these accumulations of rubbish, which are centres for all kinds of insect-vermin, but a thorough search in such places will produce handfuls of chrysalids, and thus greatly lessen the amount of the next brood of butterflies. In the winter, besides collecting the chrysalids, it is desirable at spare times to rub a strong birch-broom well up and down in the angles of the walls of sheds, or along the top of the walls beneath rough flooring, and thus make a complete clearing, before spring, of chrysalids from which the butterfly would then have hatched to start the first brood of caterpillars; the destruction of one female before laying her eggs prevents the appearance of scores of the grubs.

Hand-picking the caterpillars is a tedious remedy, but where there is no great extent of ground it is advisable as a certain cure. The application of finely-powdered lime in a caustic state, or of fresh soot, will get rid of the caterpillars, but may be objectionable with regard to after use of the vegetable.

A sprinkling of fine salt has been found very serviceable, carefully applied, so as to fall on the caterpillars; and they may also be killed by waterings of weak brine, lime-water, or soap-suds.—(M. D.)

Flour of sulphur dredged over the plants, or a weak solution of alum lightly syringed on the leaves, have not, so far as I am aware, yet been experimented with; but, looking at the success of these remedies in other cases, they are well worth a trial.

Many kinds of dressings, such as wood-ashes, &c., have sometimes succeeded, and sometimes failed so entirely that it is probable some point in the method or the time of application needs attention. It often happens that a dusting given when the dew is on, or after light watering (so as to make it adhere to the caterpillar and also to the plant), is of great service, whilst the same application given in the middle of the day is perfectly useless.

In the case of hearted Cabbages, a sprinkling of anything that would fall or wash down into the nooks of the Cabbage and lodge there, making it disagreeable to the grubs, would be of use, and for this purpose gas-lime that has been taken from the surface of a heap exposed to weather for about two months seems to answer. The lime (as tried on a small scale) does not injure the leaves, and the various "pests" infesting hearted Cabbage do not like it at all.

When seed-crops are attacked, it has been suggested as a good plan to shake the plants so as to make the caterpillars drop off, and to have a number of ducks ready to eat them as they fall. Probably if a boy was substituted, with a basket of soot or quick-lime to throw over the caterpillars, or if he trampled on them, or a ring of gas-lime was thrown round each plant to keep the caterpillars from returning up the stem, it would do much more good than the ducks. The large flocks of

ducks or poultry sometimes recommended may do much
good, but there is a direct outlay for their purchase and
food—as they need something besides the caterpillars;
they must also be tended, or they will do harm as well
as good, and altogether, except on a small scale, where
the proprietor's poultry can be turned on and benefit by
the open run and change of diet, the plan of clearing
insect-attack by this means seems rather doubtful
economy.

It has been noted that caterpillars of the Cabbage
Butterfly which appeared healthy up to a given date,
immediately after (following on sudden rain) perished,
and were found to have become mere lax skins containing
a cream-coloured fluid.—(J. C.)

Many kinds of caterpillars are attacked by purging
when feeding on wet leaves, and, looking at these points
and also that dry weather is the time when these special
pests most abound, it appears likely that a good drenching
from anything, such as a hose or garden-engine down to
a watering-pot, if nothing better was at hand, might do
much good; firstly, it would probably make many of the
caterpillars fall off, and, if treated as above mentioned
(that is, killed, or means taken to prevent their return),
many might be got rid of; and secondly, though artificial
means would not help us as much as the change of
weather, still the sudden chill from the cold water, and
the wet state of the food which would be induced if the
operation was performed in the evening, would probably
clear off many.

Good cultivation and heavy manuring of the ground,
thereby running the plants on quickly, has been found
serviceable; and the application of liquid manure will
save a crop, even when badly infested.

If, by manure or cultivation, the crop can be kept in a
state of growth that will make a larger amount of leafage
per day to each plant than the caterpillars on that plant
consume, all will be well; but if, through drought, poor
ground, or any other cause, the caterpillars take more off
than the plant makes good, necessarily it gradually

dwindles or perishes. This point is a most important one to be considered in attacks of this nature, and especially with regard to field crops, to which it is most difficult to employ any kind of insect-preventive in the shape of dressing, remuneratively.

Severe cold in winter cannot be reckoned on as a means of getting rid of the chrysalids, which is the state in which these butterflies usually pass the winter. They have been found attached to walls, and frozen so hard that they could be snapped like sticks: yet those kept for observation appeared perfectly healthy on being thawed, and produced butterflies in due time.— (J. A. R.)

During the severe winter of 1878-79, chrysalids of the Cabbage Butterflies which I had opportunity of examining appeared perfectly uninjured by cold, which ranged at various temperatures between 10° and 30° on twenty-five nights in January. The parasite maggots in a chrysalis of the Green-veined White Butterfly were also only temporarily stiffened.

The number of Butterflies is much kept down by the various kinds of small parasite flies which lay their eggs in the caterpillars or chrysalids, and especially by one kind of Ichneumon Fly (*Microgaster glomeratus*), which lays sometimes more than sixty eggs in one caterpillar of the "Large White." The maggots from these eggs feed inside on all the parts not necessary to the caterpillar's life till the time comes for it to change to the chrysalis, when, instead of turning, it dies; the Ichneumon maggots eat their way out and spin their little yellow cocoons (like small silkworm cocoons) often seen on Cabbages, from which a small four-winged fly presently appears.— (J. C.) These cocoons should *not* be destroyed.

Another kind of Ichneumon Fly, the *Pteromalus brassicæ*, figured above, lays its eggs on the chrysalis when it has just cast its caterpillar-skin, and is soft and tender. The maggots, averaging two hundred and fifty in number, eat their way into the chrysalis as soon as they hatch, and feed on its contents.

Wasps also help to keep the butterflies in check, and have been observed especially to attack the "Small White," or Turnip Butterfly.

Small White Cabbage-Butterfly. { *Pontia rapæ*, Curtis. *Pieris rapæ*, Latreille. }

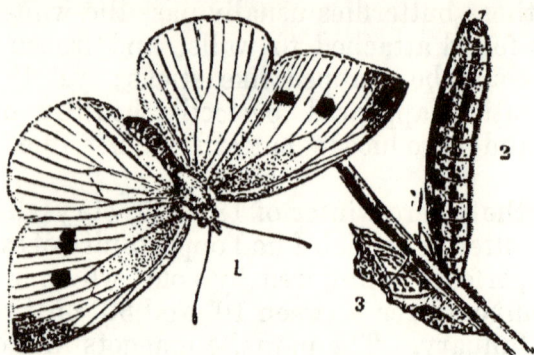

1. Female butterfly; 2, caterpillar; 3, chrysalis.

The caterpillar of this butterfly (known also as the Turnip Butterfly) feeds on Cabbage, also on Turnip, from which it takes its name of Turnip Butterfly; and also on the inner leaves of hearted Cabbage, whence the name sometimes given of Heart-Worm.

The egg is laid singly (not in clusters).

The caterpillars are green, paler green beneath, and velvety; and have a yellowish stripe along the back, and a stripe or row of spots of the same colour along each side.

The chrysalis is of a pale flesh-brown or greenish colour, freckled with black.

The fore wings of the butterfly are creamy white, with a slight grey or blackish patch at the tip (this patch not as long nor as regular in shape as in the Large Garden White). The fore wings have also one black spot above in the males, and two in the females. The under side of the fore wings is white, yellow at the tip, and has two black spots both in male and female.

The hind wings are creamy white above, with a black spot on the front edge; the under side yellow, thickly speckled with black towards the base.

PREVENTION AND REMEDIES.—For these, see the foregoing recommended for the White Cabbage Butterfly.

Green-veined White Butterfly. { *Pontia napi*, Curtis.
{ *Pieris napi*, Latreille.

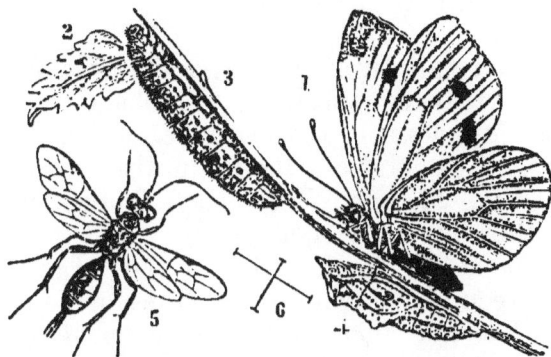

1, Female butterfly; 2, egg; 3, caterpillar; 4, chrysalis; 5 and 6, parasite Ichneumon Fly, *Hemiteles melanarius*, magnified, and nat. size.

The eggs of the "Green-veined White," which is known also as the "Rape-seed Butterfly," are laid singly under the leaves of Turnips and Cabbages, and though it is doubtful whether the caterpillars are often injurious to any serious extent, they are noted by Curtis as feeding on Turnip-leaves in 1841, and in the same year as doing much mischief to the hearted Cabbages, by gnawing into the middle of them, like the caterpillar of the Cabbage Moth.

The grub or caterpillar of this "Green-veined" butterfly is velvety; of a dull green above, brighter below, and has a row of red or reddish yellow breathing-pores along each side.

The chrysalis is pale greenish white, or yellow and freckled, and has the two ends of a brown colour.

The fore wings of the butterflies are white above, excepting at the base, which is generally black; the tip also is dusky or black, and the nerves or veins greyish; in the females the markings are blacker than in the male, and there are also two large black spots beyond the middle of the wing (the males have sometimes one spot). The under side of the fore wings is yellow at the tip, with dark veins, and two black spots. The hind wings white above, with a dusky mark on the front margin; beneath they are sulphur or pale yellow, with broad greenish margins to the nerves.

PREVENTION AND REMEDIES.—See "Large White."

Note.—The three species of Cabbage and Turnip Butterflies just described are so much alike, that it may be convenient to point out the main distinctions :—

Eggs.—The "Large White" lays its eggs in *clusters;* the two other kinds lay them *singly.*

Caterpillars.—The caterpillar of the "Large White" is bluish green above, with three lines of yellow, and is *spotted with black;* also has *tufts, or a sprinkling* of hairs. The caterpillars of the two other kinds are green, but *have no black* blotches; also they are *velvety.* These two kinds differ from each other in the "Small White" having three yellow *lines,* and the "Veined White" having a row along each side of red or reddish yellow *breathing-pores.*

Chrysalids.—The chrysalis of the "Large White" is pale greenish, spotted with black; of the "Small White" fleshy-brown, freckled with black; and of the "Green-White" pale greenish white, or yellow and freckled, with each end brown.

Butterflies.—The "Large White" usually measures about two and a half to three inches in the spread of the wings; the two other kinds are only about two inches. With regard to the markings,—in the "Large White" the patch at the tip of the fore wings is much larger, blacker, and more regularly notched on the inner side than it is in the "Small White"; also the males of the

" Large White " have no spot (or rarely have it) on the centre of the fore wings, whilst there is usually one in the case of the " Small White."

Each of the above may be easily known from the other common kind—the " Green-veined White "—by *not* having broad green veins on the under side of the hind wings.—(J. C., J. F. S., and J. O. W.)

Cabbage Fly. *Anthomyia brassicæ*, Bouché.

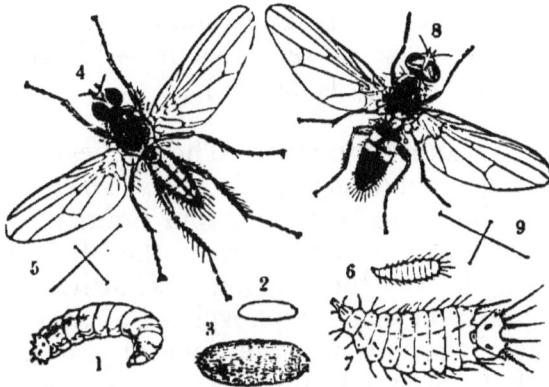

1, Larva of *A. brassicæ*; 2 and 3, pupæ, nat. size and magnified; 4. *A. radicum*, magnified; 5, nat. size; 6—9, *A. tuberosa*, larva and fly, nat. size and magnified.

CABBAGE FLY.—The maggots of this fly injure the Cabbage-crop by eating passages in the stem and roots, and sometimes destroying whole fields of Cabbage by consequent disease, or decay in wet weather, of the roots and the lower part of the stalk. They are also injurious to Turnips. The maggots may be found in hollows of the swollen Cabbage-roots. They are whitish, cylindrical, and legless, tapering to the head, and blunt at the tail, which has short teeth on the lower margin and two brown tubercles in the middle; and when full-grown they are about the third of an inch long. They then leave the plants and turn, in the earth, to pupæ (that is, their skins harden into oval red-brown cases), with a few black spots at the head, and short teeth at the tail, inside which the flies form and come out in about a fortnight or three weeks. These flies are on the wing throughout the summer, and successive generations of maggots are kept up till November; after that time the pupæ lie in the ground unhatched till spring or early summer. The fly is ashy grey, and smaller than the Onion Fly, which it much resembles. The male, however, is of a darker grey, and has a short black stripe along the back between the wings, with a curved one on each side of it, and one black stripe along the abdomen. The female is much smaller than that of the Onion Fly, otherwise is very like it.—('Farm Insects,' 'Gard. Chron. and Ag. Gazette.')

PREVENTION AND REMEDIES.—The application of lime-water has been found very serviceable. The plan adopted was soaking hot lime for twenty-four hours in water, and watering with this, when clear, in the afternoon; this was found to destroy the maggot.—(J. M'K.)

The application of superphosphate of lime is advised as a means of prevention in continental practice.

Plants that are attacked by the maggots may be easily known by their yellow or dull lead-colour, and by the leaves drooping or fading in the heat of the day; and unless the attack is checked, as above mentioned, these plants should be carefully removed and burnt, or got rid of in any way that may make sure of destroying the maggots in them; and any liquid application, such as strong brine, or ley of ashes, such as will poison them should be poured into the holes to kill any maggots that have been left.

As these maggots turn to pupæ just below the surface of the ground, it has been found a good plan to draw the earth away from the roots of the Cabbage with these pupæ or fly-cases in it, and to destroy it and them together. By this means the coming brood of flies is got rid of.

A proper rotation of crop would be one of the best methods of prevention.

The practice of taking one crop of Cabbage after another off the same ground creates a nursery for the propagation of all the insect-pests which prey on this crop; and in the case of such as turn to pupa in the earth, the grubs or caterpillars simply leave the roots, or descend from the plants into the ground, and there undergo their transformations. This may be immediately, or they may be in the earth during the winter, to appear as perfect insects in the ensuing spring or summer, and so on successively till the ground is completely infested by them.

It will thus be seen that a simple turning over of the ground, accompanied by heavy manuring, though it may be a remedy for the exhaustion of the soil, will not get rid of the grubs.

A change of crop is necessary—such as Beans, Cereals, or even Potatoes; these would not suit the Cabbage Fly for the deposit of its eggs, and planting the Cabbage afterwards on the clean uninfested land would greatly reduce the risk of attack.

Root-eating Fly. *Anthomyia radicum*, Curtis.

(For fig. of Fly, see page 31.)

The grubs of this fly are sometimes to be found in the roots of Cabbage and also in Turnips. They are of a yellowish ochre-colour, with two dark brown points at the end of the blunt tail. These grubs, and also the pupæ of this fly, are similar in shape to the grubs and pupæ of *A. brassicæ*, the Cabbage Fly (figured at 1 and 3, magnified; 2, nat. size, p. 31); the pupæ, however, differ in colour, those of the Root-eating Fly being of a paler and more ochreous colour than those of the Cabbage Fly.

The Root-eating Fly is figured at 4, p. 31, magnified; the male has an ochreous face, with a rusty stripe on the forehead; body between the wings black, with three dark stripes, and grey sides; abdomen slender, grey, with broad black stripe along the back. The female is grey,

D

with three dull stripes along the body between the wings.
—('Farm Insects.')

PREVENTION AND REMEDIES.—These flies are not noted
as causing any serious damage in this country, but may
be mentioned from their great resemblance to the
common Cabbage Fly. According to Bouché, the flies
inhabit dung by thousands in the summer, and, from
the nature of the excrementitious matter which they are
stated by him especially to frequent, it would appear
that the use of night-soil as a manure would be likely to
attract attack.

From experiments made by the Zoological-Botanical
Society of Vienna, it appears that these maggots were
injurious to crops on ground manured with bone-dust,
and also to those manured with horse-dung; but that on
ground close by, manured with superphosphate, the
plants were not at all attacked.

Anthomyia tuberosa, Curtis, figured p. 31, has been
found feeding in Potatoes, and the larvæ and pupæ have
been found where garden-ground has been long occupied
by Cabbages.—('Farm Insects.')

Cabbage Moth.   *Mamestra brassicæ*, Linn.

This is one of the garden-insects from which we suffer
regularly in the summer and autumn of each year,
sometimes slightly, but often to a serious extent. The
caterpillars do not seem to dislike the leafage of any of
our common plants, whether sweet or acrid, and they
may be found on Marigolds, Dahlias, and Geraniums, in
the flower-garden, as well as on the leaves of the Tobacco;
they frequent the Dock, amongst wild plants, and they are
sometimes found on the leaves of the Red Currant; but
we suffer most from their attacks on Cauliflower, and
on the hearted Cabbages in the autumn.

Their appetite seems insatiable (they are stated to eat
day and night), but however this may be, they soon ruin

the Cabbage by gnawing large holes down into the heart, and making what is left truly disgusting, by the excrement from the caterpillars, which remains in lumps between the leaves, or spreads downwards in wet green masses.

The moth lays her eggs on the leaves, and the caterpillars hatch in a few days and immediately begin to feed. They are usually green in their early stages, but afterwards vary much in colour, some being pale dingy green; some green, and black above; some blackish above, variegated with flesh-colour, and on the back of each ring or segment there is a short line or somewhat triangular-shaped mark, lighter at the edge, and slanting

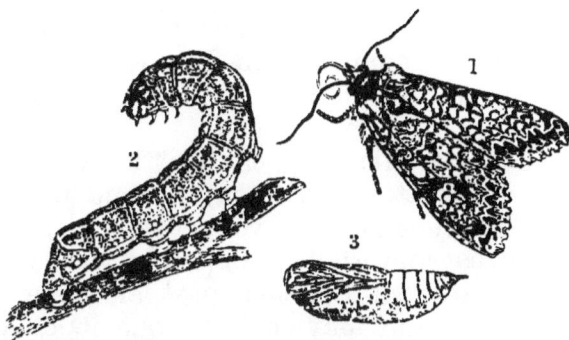

1, Moth ; 2, caterpillar; 3, chrysalis.

backward. The head is ochreous, or marbled with darker brown, and the next segment to it is black above. When full-fed these caterpillars are upwards of an inch and a quarter in length, and on being annoyed roll themselves into a tight ring. They change to chestnut-coloured chrysalids in the earth (or sometimes on the surface), and usually pass the winter in this state, protected by earth. From these the moths come out in the course of the following May or later in the summer. The moths are of a rich brown, the upper wings are variously marked (as in the figure) with black streaks and circles, and have also a large ear-shaped spot, bordered with white, and surrounded by a dark line ; the lower wings are brown, dirty white at the base.—('Farm Insects,' 'Hist. of Brit. Moths.')

PREVENTION AND REMEDIES. — Where the autumn Cabbage-crop has been much infested by the caterpillars, the chrysalids will be turned up in great numbers in the winter digging. When seen these should be immediately destroyed, or thrown into a basket to be effectually destroyed at the first leisure moment, and by this means the number of the next season's brood will be much diminished. It is no use leaving them to be killed by frost, for caterpillars and chrysalids of such kinds as have been tested at present will stand being frozen stiff without the slightest injury.

Poultry are of service in clearing the ground of these chrysalids, as the common barn-door fowls are particularly fond of them; but hand-picking is the surer method.

When the caterpillars appear on the Cabbage it is very important to attend to them, in some way or other, at once. The grubs are very voracious, and very soon—by what they eat and what they spoil by their excrement—place the attacked Cabbage past all hope. Hand-picking is of use, and may be best done by children, as their small fingers are most suitable for getting between the folds of the Cabbage-leaves, and, under proper inspection, the Cabbages may be well and rapidly cleared at a small expense. It is stated that, when nearly full-grown, the caterpillars go down by day, and feed by night; and where this is the case, what we need is something that, when once they are down, will keep them from crawling up the plant again.

Probably soot thrown close up to the stems would answer well; ashes sprinkled with spirit of tar would be useful; and so would quick-lime, as long as it remained caustic; but the application of gas-lime is an almost unfailing preservative. If this, whilst still fresh (and poisonous alike to plants and insects, if incautiously applied), is put in a narrow ring, about as wide as a finger, round but not quite close to the stem of the plant to be protected, the plant is not in the slightest degree injured, and, as far as experiment shows at present, the caterpillars will not cross the poisonous ring. The cost of the remedy is little, and in careful hands it answers well.

With regard to the hearted Cabbage, the application of gas-lime which has been exposed for about three months to atmospheric action—so as to neutralize its poisonous effects, but still not entirely to destroy the sulphury smell—has been found very serviceable. The powdered lime rolls down and lodges in all the nooks of the Cabbage, and thus makes the spots where the caterpillars most resort to, before piercing into the heart, the most distasteful to them. By comparison of gas-limed and unlimed plants in one garden, the effect has been found to be good in keeping down the grubs, without the least injury to the Cabbage. Sprinkling a good dressing of the gas-lime in this state on the surface of the Cabbage-bed is a very good preventive to attack, and acts well as a manure.

In the case of caterpillars that feed, like these, on many of our commonest weeds, it is desirable to clear away all such food-plants; and if the grubs have been noticed in summer in neglected corners, to turn over the ground in winter to see if the chrysalids are there, and, if so, to destroy them.—(' Farm Insects,' and Ed.)

Great Yellow-Underwing Moth. } *Noctua* (*Tryphæna*) *pronuba*, Linn.

The eggs of this moth are laid on many kinds of vegetables in July.

The caterpillar feeds by night; during the day it conceals itself in the ground, or under stones or clods of earth; it has been found with the caterpillar of the Heart and Dart Moth at the roots of Turnips, and also at the roots of Lettuces, and has been stated by various authors to feed on the roots of grass.

When full-grown the caterpillars are one inch and
three-quarters long, and about as thick as a swan's
quill. The head is ochreous, with two black stripes;
the general colour is very variable, from a sickly green to
a dull brown, variegated with rosy brown, and freckled
with a brown band down the back, marked with a pale
line along the middle, and a short line of black spots or
streaks along each side; the caterpillar is pale green
beneath. These caterpillars are very fat, not at all
shining, and on being disturbed coil themselves into a
ring. They feed during autumn, and may be found
during the winter months beneath the turf, or just

1, Caterpillar; 2, chrysalis; 3, moth.

beneath the surface of the ground. In spring they come
out again, and when full-fed, which may be from March
to June, they bury themselves and form earth-cases or
hollows in the ground, in which they turn to reddish
chrysalids, from which the moths come out in June or
July. They are variable in marking, but may be known
by their pale or rich umber-brown fore wings generally
mottled or spotted, as figured at 3, and by their hind
wings being orange-yellow, with a somewhat narrow
waved black band, but without any central mark. By
one or other of the three points of their large size, the
comparatively narrow black band on the hind wings, or

the absence of the dull mark on the hind wings, this moth may be distinguished from other species of Yellow-Underwings.—(E. N., J. F. S., J. C.)

PREVENTION AND REMEDIES.—The moth shelters itself amongst dry leaves and herbage, and in seasons when it is numerous many might be got rid of at a small expense by setting children to catch them with a bag-net, or merely with the hand. The moth rises before the passer-by from grass in hay-fields that have been lately cut, or in rough neglected spots; and its flight being sluggish, and itself large and very conspicuous, from its "yellow under-wings," it is easily taken. The same means are applicable for prevention that are recommended for destroying the Heart and Dart Moth, *A. exclamationis*, and most of the remedies against the Cabbage Moth, *M. brassicæ*, are also applicable to this one, but not all, as the Cabbage Moth caterpillar feeds by day as well as by night.

The caterpillar of the Yellow-Underwing feeds on Docks and on Bittersweet; consequently clearing these common weeds would be of much service. It is also said to feed on the roots of the Primrose, which should be carefully watched where this moth is prevalent.

Cabbage Powdered-wing. $\left\{\begin{array}{l} \textit{Aleyrodes proletella}, \text{ Linn.} \\ \quad\text{,,}\quad \textit{Chelidonii}, \text{ Latr.} \\ \quad\text{,,}\quad \textit{? brassicæ}, \text{ Walker.} \end{array}\right.$

This is a small four-winged Fly, very like a small powdery White Moth in its appearance, but in fact is nearly allied to the Scale Insects and to the Aphides. It may be found on different kinds of Cabbage, more or less, all the year round, and towards autumn in such enormous quantities that, if the plants are stirred, the soft white flies will rise and float in the air, and settle down like miniature snow-flakes, whence its name of the Snowy Fly. They rest for the most part beneath the leaves, and draw away the sap with their suckers; and at times do a good deal of harm. The attacked leaves may generally be known by their patchy brown or yellow state, but sometimes are entirely discoloured.

As far as is known, these Snowy Flies breed in winter as well as in summer, and are not hurt by rain or cold.

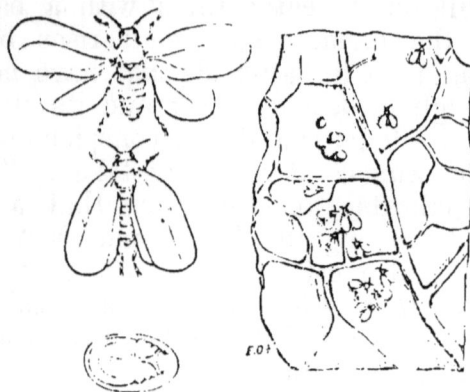

Fly and pupa, much magnified.   Flies on leaf, twice nat. size.

The female lays her eggs in a patch on a leaf; these hatch in about twelve days, and the young spread themselves on the leaf and shortly become covered with a scale, white, with two yellow spots. In this state they much resemble the common Scale Insect, and they stick fast to the leaf, which they pierce with their sucker. In about ten days they turn (beneath this scale) to a pale chrysalis, with red eyes, and in about four days more the perfect insect comes out through the scale. This is very small, only about the eighth of an inch in the spread of the wings, and is covered with a white powder. The head and the body between the wings are black, variegated with yellow; the abdomen is yellow or rosy, and the four wings are white and mealy, the upper pair having a dusky spot in the middle and towards the tip. The head has a rostrum or sucking-tube with which it draws away the juice of the leaf. — ('G. Chron.,' and 'Ag. Gazette,' 1846.)

PREVENTION AND REMEDIES.—Cabbage infested by these Snowy Flies may be known by the unhealthy look of the leaves, which are sometimes withered, but more commonly marked with whitish or yellowish patches.

These leaves should be gathered and destroyed at once. They may be thrown into a farm-yard where they will be trampled in wet manure, or into a dung-pit, and the insects thus destroyed; but it is a better plan to burn them. Throwing the leaves to a rubbish-heap is of little use, for the Cabbage keeps fresh quite long enough for the grubs to feed and turn to the perfect insect as if nothing had happened.

Probably dusting the plants with soot, or ashes sprinkled with tar, or if possible giving a good syringing with tobacco-water, or soft soap, might be of service; but if the Snowy Flies have got hold, cutting off the leaves and burning them is the best treatment.—(ED.)

Cabbage Gall Weevil. *Ceutorhynchus sulcicollis*, Stephens.

The Galls that are found singly or in clusters of roundish knots, on the underground part of the stem and sometimes on the root-fibres of the Cabbage, are caused by this Weevil.

The female lays her egg on the root, or stem, or in a small hole pierced in it by means of her long snout, and the galls form in consequence, each gall being presently found to contain in the middle of its soft cellular tissue a legless yellowish white fleshy maggot, with strongly-toothed jaws. The galls grow with the growth of the Cabbage, the maggots meanwhile feeding on the inside, till when ready for their change to pupæ they gnaw a hole through the rind with their strong jaws, and go into the earth. Here they form themselves each an earth-case. Beginning by holding a little of the material round them fast with the tip of the tail, they keep adding to it by sticking on the smallest morsels of earth, decayed wood, or pebbles that lie near, by means

of moisture from the mouth; and so gradually build up
a case round themselves, smooth inside and rough
outside, in which they turn to pupæ, and come out
again as the perfect beetle or "Cabbage Weevil" (in the
instances noted) in about two months. They seem to have
wonderful powers of endurance whilst in the grub state.
When the maggots are taken from the galls, even if not
fully grown, they will bury themselves; and if when complete their earth-cases are broken open, they will repair
them and make them as good as before.

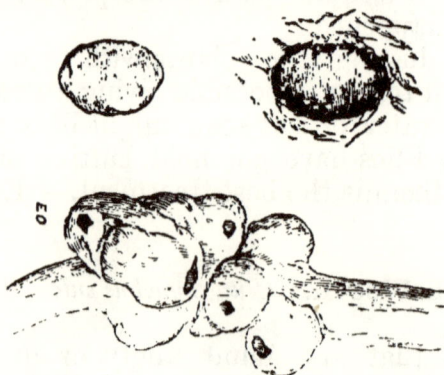

Earth-case of the larva, and case in its chamber, both magnified. Cabbage
root with galls. For fig. of Weevil see "Turnip Gall Weevil."

An amount of cold that freezes them stiff in their cells
does not do them any harm, and on being thawed and
placed on soft earth they will make their way into it and
form their earth-cases as if nothing had happened.

The perfect beetle or Cabbage Gall Weevil is about
the eighth of an inch long, blackish grey, with a long
and slender curved snout on which are placed the horns,
with a long joint at the base, so that they appear to be
elbowed; see fig. When seen under a magnifying-glass
the beetle will be found to be marked as in the figure,
with ten lines along each wing-case, and pitted with
small holes; beneath it has buff-coloured or whitish
scales.—(Ed. in 'Entomologist,' vol. x.)

PREVENTION AND REMEDIES.—Generally the galls do little harm, but occasionally run together in great masses, and exhaust the strength of the plant. Though distinct from the vegetable disease known as "Club," the galls may be found with it, and very probably may add to the unhealthy growth, though they are not the original cause.

A great deal of good might be done by burning the infested old Cabbage-stocks when drawn from the fields, instead of throwing them into rot-heaps or of digging them into the ground, as is often done. In this case the maggots are perfectly well suited for the future; they leave the galls for the earth near them, and (unless the galls are very deeply buried) in due time the beetles, with the help of their long boring beaks, will come up through the ground all well and strong; so that if Cabbage is planted over the buried stocks, it is most conveniently placed for the supply of the new brood. This may seem an overdrawn statement, but such treatment and results are only too true.

In common garden cultivation the old plants might easily be got rid of by throwing them to the pigs, or into a farm-yard where the grubs (if they came out) would be choked in wet manure. In field cultivation, where the stocks are cleared by the cart-load, it would be desirable to burn them, or shoot them into a field-pond away from houses where the smell would not cause annoyance, although it would hardly be worse in this way than when the old stocks lie rotting in heaps. But whatever is done, it should be done at once, before the maggots have time to leave the galls; and if this simple matter of prevention was properly attended to, most likely very little would be needed by way of remedy.

When seedlings are being planted, all with galls on them should be rejected, or the galls scraped off and destroyed. Change of crop is very important. These weevils attack the Cabbage and Turnip, but not other farm crops, such as Carrots, Parsnips, Corn, &c., and by the time that the Cabbage or Turnip rotation comes round again, the ground would be free from the weevils and their maggots.

Gas-lime, sown broadcast and then pointed in, has been found of service in getting ground intended for Cabbage-crops clear of this maggot.—('Ag. Gazette,' 1879.) Gas-lime is also serviceable as a dressing, accompanied by deep trenching; so is caustic lime, or soot.

Wood-ashes are also a good preventive for insect-attack on the roots, and are otherwise very beneficial to the growth of the plant.—(M. D.)

A dressing of marl or fresh soil is very serviceable on old garden ground. It is also of great use to trench the ground so as to bury the earth-cases well down, beyond the power of the weevil when it develops to come up again.

In some Cabbage-growing districts, spent Hops are largely used as a manure, and are stated to be an excellent preventive against the attack of this and many other pernicious insects.

CARROT.

Carrot Fly. *Psila rosæ*, Fab.

1, 2, and 3, Larvæ, nat. size and magnified; 4, infested Carrot; 5 and 6, pupæ; 7 and 8, Carrot Fly, nat. size and magnified.

The damage done by this "Worm" or maggot is known as "rust," from the peculiar reddish or rusty colour to which the gnawed parts turn.

The maggot is legless, white or yellowish, and shiny; about a quarter of an inch long, pointed at the head, and cut short off at the tail. The head is a black horny substance which contains the mouth, the tip of which is like a hook; the tail has two little black tubercles. The grubs may be found in winter as well as summer, and attack all parts of the Carrot-root by gnawing galleries on the surface, or into the substance of the root; but whilst the roots are young, the grubs appear generally to attack the lowest part. If infested Carrots are carefully drawn from the earth, the grubs will be seen on the root, sticking out of their burrows by about half their length. The attacked Carrots may be known by the outer leaves turning yellow and withering, while the roots gradually sicken and die from the injury to the fleshy part, the growth of the root-fibre being also often completely destroyed.

When full-fed, the maggots leave the Carrots and turn to pupæ in the earth. These pupæ or fly-cases are shiny, of a rusty or ochre colour, pale russet at the ends, with two little black points at the tail. The fly comes out in three or four weeks in summer, but in winter the pupæ remain unchanged, and the fly does not come from them till the following spring or summer. It is very small, less than half an inch in the spread of the wings, blackish green, with a round rusty ochre-coloured head, and ochre-coloured legs. The abdomen is sharply pointed in the female, and the two wings are iridescent, with bright ochre-coloured veins.—(' Farm Insects.')

PREVENTION AND REMEDIES. — The following notes regarding Carrot-cultivation will be found to bear in various ways, suitable to different circumstances of soil and climate, on the main points of—1st, such preparation of the ground in autumn or winter as will ensure favourable conditions for a healthy, vigorous, and uninterrupted growth from the first sprouting of the seed; 2nd, thinning at such a stage of growth, in such circumstances of damp weather, or with such watering or treatment after thinning as may least expose the plants to the attack of the Carrot Fly, which frequently occurs after this operation. Whether the fly is attracted by the scent of the bruised plants, or what brings it is not clear, but it is very clear that as it goes down into the ground to lay its eggs on or by the Carrots, that all operations which leave the soil unusually loose and open, lay at the same time the Carrot-roots open to attack; and it will be observed that the various methods of treatment, in regard to thinning, bear on the means of meeting this difficulty. Observations as to kinds of manure, and special applications, as salt, gas-lime, wood-ashes, &c., found to be serviceable, are also given.

The notes are classed as far as possible in groups, with reference to the point mainly brought forward in the observation.

At a locality near Dingwall, N.B., where the ground is

a damp heavy clay, and consequently unsuitable for Carrots, a piece was prepared in 1874, on which they have been found to answer well subsequently year by year. The soil was removed to the depth of two feet, and replaced by a mixture of well-decayed leaf-mould, sand, charcoal, soot, and light soil; on this the Carrots have only suffered slight injury, when other beds in the neighbourhood have been destroyed by the fly.

It is advised that Carrots should be thinned out to the distance at which they are to be grown, when they are weeded the first time, for if they are thinned after they come to any size, the soil is opened round the Carrots, and if dry weather follows, the fly is sure to attack them. —G. M'K.

At another locality near Dingwall it is the custom to sow sparingly, and not thin the Carrots till fit for use; and—in illustration of attack of grub or "worm" following on thinning—it is observed that in 1880, after commencing this process with the second sowing of Carrots, "the grub commenced too, and within three weeks spoiled them for use. Late ones alongside escaped until we began to use them; but, by keeping to the side the grub was on, it kept up, but did not advance beyond the damaged part."—(A. S.)

"Unless the Carrots are thinned very early it should not be done till they are fit for use, as there is great danger of attracting the fly by the broken pieces of root remaining in the ground."—(A. A.)

At Dunrobin, N. B., the ground is carefully trenched, and the manure kept about a foot or fifteen inches from the surface; and it is the practice to use the best seed, sow thinly, and thin early, or not till the Carrots are fit for use. "If Carrots require thinning, this should be done when an inch or two high; if thinned after they are from two to six or eight inches high, the fly seems to attack them more readily."—(D. M.)

On ground near Kirkwall, in the Orkneys, where the soil was of a strong clayey nature, unsuitable for Carrots, it was found, after all previous experiments had failed to produce a good crop, that (instead of using ordinary manure) digging in about the same quantity of peat and sand, and giving a good watering with gas-water before sowing, was quite successful.—(T. M'D.)

In a locality where the Carrot-crop suffered severely from the fly on the ordinary soil, a piece of peaty ground was taken for Carrot-growing which produced good crops yearly.—(J. D.)

In the following notices the soil appears to have suited the Carrots perfectly, and the attack of fly to have followed on thinning :—

At Lockerbie, N. B., where change of soil was tried as a remedy for fly, a portion of ground was cleared out to the depth of twenty-two inches, and the vacancy filled in with a compost of four parts good brown peat, four parts light fine yellow loam from Vinery border, two parts well-decayed leaf-mould, and two parts river-sand ; the whole being thoroughly mixed, and no manure being used. The Carrots sown on this were thinned at the end of May, and did well till the 17th of June, when the fly was found to have begun its ravages.—(F. G. F.)

On a stiff clayey loam near Isleworth the same experiment was tried on a smaller scale, by removing soil to the depth of about eighteen inches, and filling in the space with a mixture of about one-third decayed leaf-mould of Elm-leaves, one-third of the loam that had been dug out, and one-third of mixed coal-ashes, broken peat, and a little white sand. The Carrots did well on this till the end of June, when they were thinned, and the grub shortly after appeared amongst those most disturbed by the thinning. Copious watering with an occasional application of dilute Soluble Phenyle stopped the attack, and threw the Carrots into vigorous growth.—(Ed.)

The following notes have reference to methods of cultivation of the ordinary garden ground, to the addition of *salt*, and other occasional applications :—

At Skibo, N. B., the course is adopted of having the ground double-dug before hard frost sets in, the manure being put at the bottom of the trench. When complete, a good dressing of salt is given, which washes down before spring, and previous to sowing a good dressing of soot and wood-ashes is applied. Care is taken to perform the first thinning as soon as the plants can be handled, leaving them an inch or two apart. When the thinning is complete a sprinkling of guano is given, and a thorough watering, if the weather is dry. Particular attention is paid to keeping the Carrots growing without any check till some time after they are finally thinned, when, if all has gone well previously, they are considered safe. Still, however, it is thought well to look them over occasionally, and if any are found drooping, to have them pulled up and burned. This method of cultivation has been found, after many years' experience, to answer well. —(A. F.)

Near Berridale, N. B., the method adopted is to turn up the ground in winter as roughly as possible, and give a good dressing of salt. In spring, before sowing, it is forked over, and a dressing given of peat and ashes. When the Carrots are thinned, a mixture of soot and water is poured over them; paraffin is also used, in the proportion of an English pint of paraffin to two gallons of water, but care is recommended in the use of it in dry weather, lest it should burn the plants.—(J. S.)

Frequent waterings of salt water are mentioned as of service to an attacked crop.—(T. M'D.)

At Dalkeith, N. B., good Carrots, tolerably free from injury by the fly, were grown on ground which had been previously heavily salted for Asparagus. This was upon soil which had been used for a kitchen-garden for forty years, and upon which endeavours had failed to get a good crop in any other way.—(M. D.)

At Ballinacourte, Tipperary, the most severe portion of an attack of the Carrot Fly occurred in the centre of a field which had not received the same treatment as the part round it, and the plant was consequently more feeble. In this case bad manure and the absence of salt were considered to be the cause of the attack. It is advised that good manure—rich and well-rotted—should be applied at the autumn ploughing, and artificial manure, with salt, applied in the drill at sowing-time.— (D. S. S.)

In the following notes mention is made of the use of *paraffin* in regular course of cultivation, or as a dressing in case of attack, proving a very good means of prevention :—

The crops have usually proved good at Oxenford Castle, N. B., on a bed of tree-leaves covered with soil composed of emptyings of flower-pots, boxes, &c., and consequently of a light friable nature. In 1879, however, the fly proved destructive, and in 1880, in order to experiment, the beds were beaten firm after the seed was sown, and lightly covered with soil, as above mentioned. A good dressing was then given with wood-ashes in which paraffin oil had been mixed in the proportion of one quart to a barrow-load of ashes (about one hundred-weight). When the plants were about four inches high, a second dressing of the same mixture was given, and a thoroughly excellent and luxuriant crop resulted. It is observed that mixing the paraffin with some absorbing substance is better than simply watering with it in a dilute form, as it lasts longer, and is more gradually carried down into the soil.—(A. A.)

Sand saturated with paraffin oil and strewed amongst the Carrots, and afterwards watered in, is noticed as a good preventive.—(G. M'K.)

At Hopetoun, N. B., the Carrot-crop was not satisfactory till the use of paraffin oil (in the proportion of two wine-glassfuls to a gallon of water) run along the drills after thinning was tried, and proved very successful. —(D. M'L.)

Watering with paraffin oil in the proportion of a wine-glassful to a gallon of water has been found to be of service.—(R. S.)

It is also noted that, in an attack of the Carrot Worm near Guildford, Surrey, the patches to which a dressing of paraffin and soot were applied turned out well, and yielded good roots.

The following notes refer more especially to the successful use of *gas-lime*, *wood-ashes*, and *soap-suds*, and also to successful cultivation on ground which has a liberal supply of rich manure thoroughly incorporated with it by the cultivation of the previous crop :—

It is mentioned that, at Marchmont, N. B., the system adopted is deep trenching after Celery, and cropping with Carrots without addition of manure, and as long as this practice is adhered to, the fly rarely appears.—(P. L.)

At Dalkeith, a dressing of gas-lime forked into the soil before sowing is found to be a good remedy, as also is an application of ammoniacal liquor or strong liquid manure, whenever the Carrot Fly is first noticed on the wing.—(M. D.)

At Callander, N. B., where the grub has been found troublesome, the use of gas-lime was tried with success. The ground was rough-dug at the beginning of winter, and the gas-lime was sprinkled over the ground till it was white, and then pointed-in about four inches. The Carrots on ground so treated escaped all attack from fly, whilst those on another piece not so treated were destroyed; and a second sowing made to replace these on ground dressed with gas-lime also did well.—(T. B.)

Gas-lime dug two or three inches thick into the bed before sowing, is found to answer at Hopetoun; but it is observed that perhaps the best and cheapest cure for the worm in Carrots is a solution of alum in water, applied by a watering-can with a rose. This is mentioned as perhaps even better than paraffin, as there is less risk of misapplication.—(J. M'L.)

A note of treatment, in 1879, at Gordon Castle, Fochabers, points to the effect of wood-ashes in preventing attacks so far down the root as the alkali is carried in solution by the rain. Some of the ground was trenched, other portions sown without trenching, but the whole otherwise treated alike,—that is, by covering over the seeds in the drills with fully half an inch thick of sifted wood-ashes. The Carrots were all attacked by the grub, and proved an entire failure : but it is noted the

attack was about two and a half to three and a half
inches below the surface, the upper part of the Carrots
remaining perfectly sound.—(J. W.)

Wood-ashes are noted as applied in successful treatment, after the ridges are levelled, in a dressing not less
than half an inch all over the ground; this is dug-in one
spit deep (the ground being made level at the same time),
and soot well dusted over the surface. The ground is
then ready for the seed.—(A. I.)

At Torloisk, in the Island of Mull, the ground was
trenched about two feet deep, and a good layer of farm-
manure placed at the bottom of the trench in winter, and
prepared in the usual way in the spring; but before
sowing, deep holes were made about eight inches apart,
and filled with a compost of soil from potting-shed, soot,
pigeon-manure, lime-rubbish, sea-sand, and wood-ashes,
and a few seeds were placed in each hole. The result
was an excellent crop of clean Carrots; but a few lines
tried without the holes, but with a little of the compost
in lines beneath the seed, were destroyed—(A. G.)

The simple application of frequent waterings has a
very good effect in keeping down "rust" or attack of
Carrot grub in dry weather. The water keeps the
ground fairly compacted against entrance of the fly for
oviposition, and also keeps the plants from falling into
the checked and stunted condition in which they are
particularly liable to attack; and if given with a rose
over the foliage, the watering has the further advantage
of clearing out, for the time being, all flies which are
harbouring there, and which at once take wing before the
falling shower.

One more remedy remains to be named, which has
been recommended for such a long course of years that
presumably it is of service. This is, to prepare ground
for Carrots by a dressing of spirits of tar mixed with
sand. One gallon of spirits of tar well mixed with a
barrowful of sand and then hand-strewed over the
ground, is enough to dress sixty or seventy square yards.
This dressing may be applied in the autumn and dug-in,

or after the Carrots are sown, or it may be strewed at the time of sowing.

The above observations give a variety of means, all tending to the same results of pushing on healthy uninterrupted growth, with such subsequent treatment as will either not induce attack of the fly, or is likely to counteract attack, if it occurs.

To these methods of cultivation a note may be added regarding treatment of ground on which there has been a badly-infested crop of Carrots:—

Although the summer broods hatch in three or four weeks, the maggots may be found in the roots during winter, and they change to pupæ in the earth adjacent. It is therefore very desirable that all infested Carrot-beds should be thoroughly cleared of roots in the autumn, and the ground well dug, or trenched, so that such maggots or pupæ as remain in the bed may be destroyed; some may escape, but the larger number will thus be buried too deeply to come up again, or be thrown on the surface to the birds; and a dressing of gas-lime will be serviceable in destroying such of the grubs as are lying near the surface.

Common Flat-body Moth. *Depressaria cicutella,* Curtis.

This little moth is one of three nearly-allied kinds, of which the caterpillars injure our Carrots, and sometimes also our Parsnip-crops. I only give a figure of one kind, as they are much alike in shape. They may be known by the rather long and narrow upper wings being laid flat one over the other when at rest, and by the abdomen being flat or depressed, whence the common name of Flat-body Moth, and the scientific name of *Depressaria.*

The caterpillar of the moth described above feeds on the Carrot-leaves, which it cuts so that it can bend and roll them up into small cylinders spun together by its threads; each end is left open, so that when alarmed it can lower itself to the ground by a thread spun from its mouth.

The caterpillar is half an inch long, grass-green, with a darker green line along the back and along each side; it has ten warty black spots on each ring, a brown head, with two brighter brown spots, and the back of the segment behind the head is brown, with a black margin.

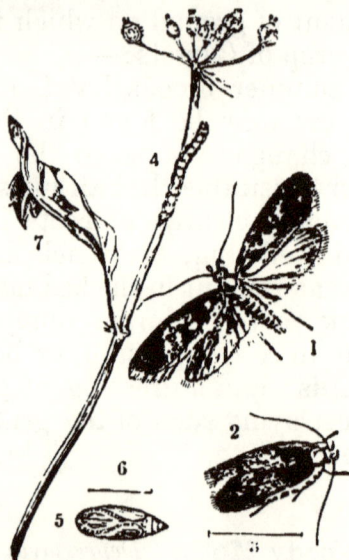

Common Flat-body Moth, caterpillar, and chrysalis.

When ready to change, the caterpillars become rosy beneath, and then turn to chrysalids of a deep yellowish brown, sometimes in a folded leaf, sometimes in cocoons in the earth.

The moth is rather more than three-quarters of an inch in the spread of the wings, shiny like satin, of a pale reddish ochre-colour, the upper wings freckled with brown and black, and with two or three white dots with dark edges towards the middle; the under wings are yellowish grey, satiny, and fringed.

There are two broods. The June caterpillars come out as moths in August; those found at the beginning of September come out as moths (which live through the winter) at the end of November.—(From 'Farm Insects.')

PREVENTION AND REMEDIES.—I am not aware that the caterpillar causes serious damage in England. Should it be troublesome, probably shaking the Carrot-leaves with a stick, or in any other convenient way, and throwing soot, or lime, or anything that would injure the caterpillars and keep them from returning to the leaves, would be serviceable; they fall on being disturbed.

Their natural enemies are the small solitary Wasps (*Odyneri*) which may often be seen in summer, and are distinguishable from the common Wasps by being generally smaller, and by their large heads, and the broad stripes of black on the sharply-pointed abdomen. These small Wasps make their burrows in sand-banks, bramble-stems, decayed posts, and the like places, in which they collect caterpillars for the food of their own larvæ.

Carrot-blossom Moth. *Depressaria daucella*, Curtis.

The caterpillars of this small Moth sometimes do much harm to the Carrot seed-crop in July and August.

They draw the tips of the flowering head (the umbel) together with their webs, and inside this chamber they feed on the flowers and seeds.

The caterpillar is greenish grey, or yellowish, with black hairy warts, and some faint streaks along the back; and the head, as well as the upper side of the first segment behind it, is brown or black; it is only about half an inch long when full-grown.

Sometimes the caterpillar changes to the chrysalis in the flower-head; sometimes it bores for this purpose into the stem.

The moth is little more than three-quarters of an inch in the spread of the upper wings. The head and body

between the wings are reddish brown, freckled with
black; the upper wings are of the same colour, freckled
with white, and having black streaks, and the under side
is dark; the hind wings are light grey.

PREVENTION AND REMEDIES.—The caterpillars fall down
by a thread when disturbed, therefore shaking the
Carrot-tops and destroying the caterpillars that fall is a
good remedy; but it should be remembered that unless
the grub is immediately destroyed, or its return prevented, it will go back again, and no good be done by
the shaking.

Powdered hellebore is suggested as being a good thing
to dust over the infested plants, applied when the dew
is on.

It is stated by various writers that this moth much
prefers the Parsnip to the Carrot for deposit of her eggs,
and consequently by planting Parsnips about eight feet
apart amongst the Carrots, the latter will be saved from
attack, and also the Parsnip-tops, with the caterpillars
and chrysalids thus collected, may be conveniently
gathered and burnt. This, however, needs a deal of
care, or the caterpillars will drop down by their threads
and escape. Breaking the heads off over a tub is
suggested as a good remedy by John Curtis in 'Farm
Insects,' from which the above notes regarding Carrot
Moths have chiefly been taken; but it would be well in
this case to have a thick mixture of soot and water, or
quick-lime, or anything that would destroy them at once
or prevent their escape, placed at the bottom of the tub,
rather than to depend on burning.

Purple Carrot-seed Moth.   *Depressaria depressella*, Curtis.

The caterpillars of this Moth resemble, both in habits
and appearance, those of the Carrot-blossom Moth, but
are rather smaller, being hardly more than a quarter of
an inch long. They are brownish grey in colour, and

the black hairs grow from white instead of from black
warts. The sides of the body have swollen edges.

They feed (in company) in the flower-heads of the
Carrot, and there some of them change to chrysalids in
a light grey web; as in the case of the Carrot-blossom
Moth this habit is not constant, for in the autumn they
eat into the stalks, and there change and winter. At
the first mild weather the moths come out, but shelter
themselves in any convenient spot on the return of cold.

The male moth is about half an inch in the spread of
the wings, and has the head and body between the wings
ochre-coloured; the upper wings chestnut, with pale
ochre scales sometimes placed in patches; the tail is
blunt.

The female is rather wider in the spread of the wings;
the colour is similar, but the chestnut of the upper
wings is brighter, and the scales are whitish and form a
kind of oval mark; also the tail is pointed.—('Farm
Insects.')

PREVENTION AND REMEDIES.—Their habits and food-
plants appear like those of the Carrot-blossom Moth—
which see for means of prevention, &c.

CELERY.

Celery Leaf-miner. *Tephritis onopordinis*, Curtis.

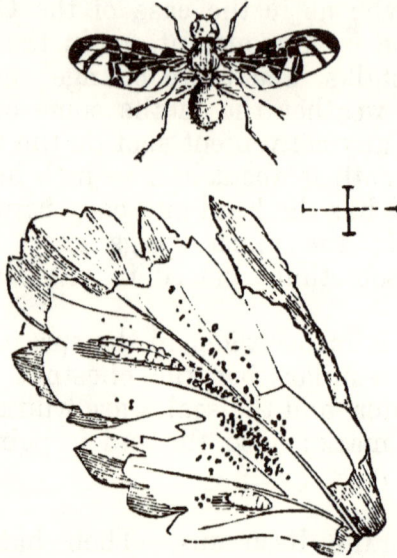

Celery Fly, magnified; line, showing nat. size; larva and pupa figured
in blistered leaf.

The Celery Fly lays her eggs on, or in, the Celery-leaf,
and from these there hatch maggots much like the Onion
Maggot (of which see fig.) in appearance. These
maggots are fleshy, legless, pointed at the head, and
blunt at the tail, white or pale greenish in colour.
They feed between the upper and under sides of the
Celery-leaf and, by eating away the substance, cause
large blister-like patches, which are white at first, and
turn brown as the skin dries. Where there are many of
these blisters, the leaves are destroyed, and the plants
are consequently destroyed or much injured.

When the maggots are full-fed, the skin hardens, and
they turn to brown or ochry oval pupæ in the leaf or in
the earth; generally they leave the leaf and turn to
pupæ in the earth.

The "miners" go through their changes from the egg to the perfect fly so rapidly as to give time for two or more broods during the year. The fly is rather more than the eighth of an inch long, and about three-eighths of an inch across in the spread of the wings. The colour is ochreous or brown, the eyes deep green, and the two wings are transparent and mottled with patches of brown. The poisers (that is, processes like a knob on a short stalk, which project from the body and take the place of the second pair of wings in the *Diptera*, or two-winged flies) are ochre-coloured.

They may be found in great numbers in the middle of May, and grub-blistered leaves may be found from the time the Celery is planted out till Christmas.

The maggots of the autumn brood turn to pupæ, and remain in the ground in that state during the winter.

PREVENTION AND REMEDIES.—A great deal might be done in the way of prevention by attending to the ground on which infested crops have been grown (and which is almost certainly full of the pupæ), and also by taking care to have infested leaves, or refuse leaves with grubs in them, destroyed at once. The plan of throwing them on the rubbish-heap is especially bad in the case of this maggot. It has great powers of endurance; I find, by experiment, it will stand damp and mouldy surroundings, or, as a pupa, the extreme opposite of very dry ones, and come out in due time as the fly in perfect health.

The only certain way of destroying the maggots in the leaves is to burn them.

With regard to treatment of the ground :—The greater part of the maggots go down into the earth to change to pupæ, that is, to the little brown cases, out of which the fly will come up presently ; these pupæ may be found in multitudes if the soil is turned up. It would be a good plan to skim off about three or four inches depth of soil where there has been a bad attack, and burn it (if only a small quantity), or get rid of it thoroughly, with its contents, in any other way that may be preferred.

If this process cannot be carried out, the ground might be trenched, care being taken to turn the top spit into the bottom of the preceding trench, and so bury the pupæ too deep for them to develop and come up again as flies. Rough-digging might be of benefit, by placing some of the fly-cases within the reach of birds, and burying others in the same manner as in trenching. A more certain plan, however, would be to give the ground a dressing of fresh gas-lime, which, on being "pointed-in," or mixed with three or four inches of the surface-soil, would destroy the pupæ; and sufficient time would elapse before the next crop was sown for the gas-lime to have gone through the chemical changes which turn it to a valuable manure. (See ref. to gas-lime in Index.)

The plan commonly recommended to get rid of the maggots is to pinch them in the blisters, and this is good so far as checking further attack from the next brood of flies which would have come from these maggots is concerned; but, with regard to the attacked plants, unless the pinching is very carefully done, it causes nearly as much damage to each leaf as the maggot itself. If it is simply crushed, it is all very well, but if a piece of the leaf is torn, or cut out, the damage is great from the operation.

Sprinkling the leaves with a good coat of anything that is disagreeable to the Celery Flies (and so prevents them from egg-laying), and that will promote the healthy growth of the plants, is certain to be of use.

Soot has been found serviceable for this purpose, and also a mixture of one part of unslaked lime, one part of gas-lime a month from the works, and two of mixed dry earth and soot, all well stirred together, and scattered liberally on the plants and the ground. A good start is of great importance in this case, and a check at planting out in the trenches particularly to be avoided. The damage is from the plant being weakened by loss of leafage faster than it can make new growth to replace it, and any amount of manure or water, or treatment of any kind that will run on healthy growth rapidly, will be of service.

Celery being a plant that naturally grows in wet places, suffers from drought, and the Celery Fly thrives in dry weather. For this reason it is desirable, where practicable, to turn a good supply of water into the trenches from time to time. This makes the manure at the roots available, and the evaporation keeps a damp air round the plants very good for growth.

It is also of service to start with the garden-engine, or hose from water-supply, or large-rosed watering-can, at one end of the row, and send a good power of water strongly at the leafage, going regularly forward from one end of the row to the other. This clears off swarms of insect-vermin, which may be seen preceding the operator down the row; and if, whilst the leafage is still moist, any dressing that may be known to be serviceable is sprinkled over the leaves, it will adhere for a while, and be very useful in keeping off attack.

Celery-stem Fly. *Piophila Apii*, Westwood.

The maggot of this Fly may be found in winter and early spring inside the solid part of the Celery-stem, and in the stalks of the leaves. The attack seems to begin at the lowest part, and gradually to be extended upwards; and may be traced by the worm-eaten passages, which turn of a rusty-red colour on their gnawed surfaces. The maggot is legless, cylindrical, and yellowish white; and, like many other kinds of maggots of the two-winged flies, has the hinder end of the body bluntly rounded, and the front end, when stretched out, shows a black horny apparatus (in this case of two hooks) by which the maggot cuts its food from the soft substance of the plant, and which, with the head containing it, can be withdrawn into the body at pleasure.

When full-fed the skin of the maggot hardens, the fly forms inside, and the spring brood of flies comes out in May—possibly there is a second brood later in the year.

The head of the fly is chestnut-colour, black above; the body, including the abdomen, glossy black, sprinkled with fine hairs of a golden grey; the two wings are colourless, excepting the yellow veins; the legs straw-colour, with feet of a more dusky tint. The size of the fly is about the same as that of the Celery Leaf-miner.— (J. O. W. in 'Gard. Chron. and Ag. Gazette,' 1848.)

PREVENTION AND REMEDIES. — The only method of prevention that seems available is to destroy all worm-eaten stems. These may probably have the grubs still in them, and, if only thrown on the rubbish-heap, these larvæ would turn to pupæ, and the flies come out presently as well as if they had not been meddled with. If this rubbish is burnt, it is the safest course; but any plan that buries the grub too deeply for the fly from it to come up again, or secures that it is destroyed, will do equally well.

CORN.

Grain Aphis. { *Aphis granaria*, Kirby ; Curtis.
{ *Siphonophora granaria*, Buckton.

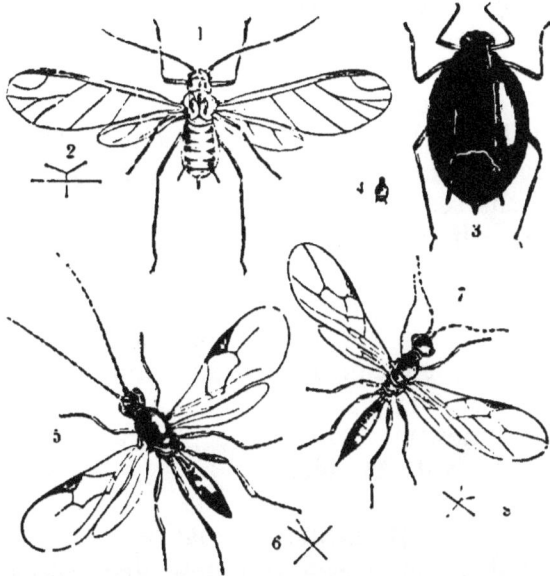

1—4, Aphides, winged and wingless, nat. size and magnified; 5 and 6,
Aphidius avenæ; 7 and 8, *Ephedrus plagiator* (parasite flies) nat. size and
magnified.

This Aphis is to be found on Wheat, Barley, Oats, and
Rye, and is sometimes very hurtful.

Early in the summer the Aphides may be found
sucking sap from the leaves or stems of the young
plants, but later they attack the ears, inserting their
suckers close to where the grains spring from the
central stalk of the head. It is stated that the young
Aphides mostly attack the upper side of the leaf-blades,
and the winged female the ear. Sometimes as many
Aphides are found as there are grains in the ear,
sometimes the ear is choked with them, more than
two hundred having been counted at once.

The injury is caused by the plant-louse inserting its

beak or sucker, and thus exhausting the plant by drawing away the sap, and also causing an irritation (by means of the many punctures) that is injurious to the plant-growth. When the ear begins to harden so that the plant-louse cannot drive the sucker in, it is safe. The fig. shows a winged Aphis, and the dark colour of those from which parasites have hatched.

The wingless viviparous females are described as green or brownish green, with brown horns, and the horn-like tubes at the end of the abdomen also brown; the eyes red. The winged females have the abdomen green, and the rest of the general colour pale brown, or rusty yellow; the tubes on the abdomen black; eyes red. The wings are green at the base, with brown veins, and the spread of the wings is about a quarter of an inch. In its earlier stages the plant-louse is green; the pupa has a more golden tint in autumn, and its wing-cases are pale brown.

It is not known where the eggs are laid.—(Mainly from 'Mon. of Brit. Aphides.')

PREVENTION AND REMEDIES.—Dusting with lime, whilst the Corn is still young and the plant-lice spread about on it, has been suggested as useful; and soot would be likely to do good, it being a fertilizer, as well as disagreeable to the insects; but probably a dressing of any kind of manure that would encourage healthy growth would be best. The nature of this would depend on the soil, and state of the crop.

It has been noted that, when the crop was late, it suffered severely; when early, it suffered less; whether this is always the case does not appear, but it is likely, partly from the leaves and ears having passed the soft delicate state in which the plant-louse can pierce them with its sucker, before the insect appeared to attack them, and partly from the larger and more vigorous plants being better able to bear attack than the young ones.

In a leading article of the 'Gardener's Chronicle and

Agricultural Gazette' for August 15th, 1868, it is remarked:—"The plant-louse in the hot summer of 1864 was, and is again during the present hot season, a great pest. It mostly affects the later Wheats, which in some cases are covered with hundreds of these creatures, which can only live by sucking out the juices from the ear; hence early Wheats are mostly too dry to be injured at the time when the creatures are developed in sufficient numbers to be mischievous."

In an extremely bad outbreak of this Aphis on one hundred and ten acres of Wheat, in Cheshire, during the wet summer of 1879, the Wheat was March-sown, and was about six weeks later than in average seasons; and in the middle of September every *green* head in the field was blasted by them.—(S. L.)

In another instance recorded the infested ears were noticed as tapering from the middle upwards, and it was found the upper half was mainly infested by the Aphides, this apparently from being rather less advanced, and consequently softer.

The various observations point to any method of cultivation or date of sowing being desirable that will get the plant on in good time, ahead of the main appearance of this kind of Aphis.

How this Aphis passes the winter is not clearly known. Mr. Walker states that it migrates in autumn, from Wheat to several kinds of grass. It certainly infests several of our commonest grasses. It is found on the well-known common grasses, Rough Cock's-foot, Wild Oat, Soft Brome, Wall Barley-grass, also on Soft Grass (*Holcus*), Meadow Grass (*Poa*); and though it has not hitherto been noted as found during winter at the roots, yet such may turn out to be the case. However this may be as regards the roots, the rough grasses that give it support on their tops would be better away. It is very desirable to encourage all the small Aphis-feeding birds, especially Titmice, so far as they are not injurious also to the Corn.—(From 'Farm Insects,' 'Notes by Observers of Injurious Insects,' &c.)

F

Daddy Long-legs. } *Tipula oleracea*, Linn.
Crane Fly.

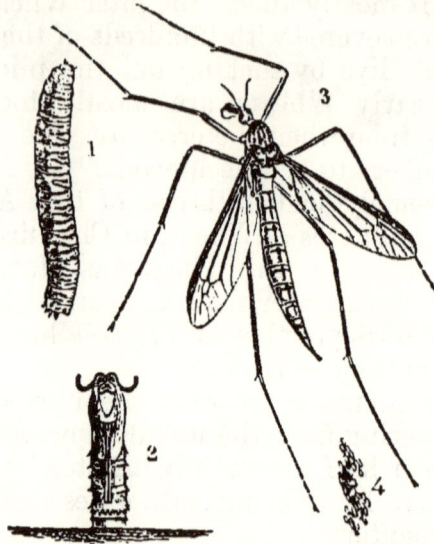

1, Larva; 2, pupa-case standing up in the ground; 3. fly; 4, eggs.

The Flies commonly known by the name of Daddy Long-legs, or Crane Flies (which produce "the grub" known also as "Leather Jacket"), are to be seen in multitudes, especially in autumn, in neglected grassy spots, meadows, Clover-leas, and on marshy ground, where they deposit their eggs, and are the cause of enormous damage to Corn and Turnip-crops, in consequence of "the grub" gnawing the young plant just below the surface of the ground, and thereby destroying or lessening the crop very much.

The female lays her *eggs* (mainly during autumn) in the ground, or on the surface, or on damp grass or leafage close to the surface of the ground. These *eggs* are small, black, and shiny, so small and so numerous that as many as three hundred are to be found in one female, forming a mass which occupies nearly the whole of the abdomen.

The *maggots* or *grubs* hatched from these eggs (to which the name of "*the grub*" is especially applied) are cylindrical, legless, of a dirty greyish or brown colour, wrinkled across, and when full-grown about an inch to an inch and a half in length.

The tail of the grub is cut short off, and tubercled; the head is protruded as a blunt point, armed with two strong black jaws. When the grub is in movement, by this pointed shape and its great power of contracting and expanding in length, it can pierce the ground, or wriggle itself forward without the help of legs. The grubs are to be found at night on the surface of the ground. From the toughness of their skins, these grubs are sometimes known as "Leather Jackets," and when full-grown are of the shape and size shown at fig. 1.

Daddy Long-legs grubs may be found as early as February (as during last year, 1880, when they were destroying hundreds of acres of autumn-sown Wheat on heavy land after Clover, round York at that date). The work of destruction, however, usually commences later,— about the beginning or middle of May,—and continues till the end of June, or even to the beginning of August. Much will depend upon the state of the weather, the character of the soil, and the condition of the plant at the date of attack, as to the amount of damage which the farmer will have to sustain.

The flies and grubs may be found throughout the summer, but for the most part the grubs change to the pupa state, in which they can do no further harm so far as eating is concerned, from July to September.

They change under ground, or under the protection of weeds; the *pupa* (or *fly-case*) is furnished with stout spines, pointing backwards, by means of which it can raise itself upward through the ground. Very soon the horny covering splits, and leaving the empty case (fig. 2) standing upright, with the hole in it through which it has escaped clearly shown along the back (as figured), the Crane Fly or Daddy Long-legs comes forth, spreads its legs and two wings, and appears as seen at fig. 3.

The *males* first appear in August.

Besides the species known as *Tipula oleracea*, of which the tawny brownish appearance is well known, there is a smaller kind, of a yellow colour, spotted with black, known as *T. maculosa*, or the Spotted Crane Fly, and another larger kind, known as *T. paludosa*, or the Marsh Crane Fly. These two kinds are hurtful to the crops in the same manner as the common Daddy Long-legs, and require the same methods of prevention.

PREVENTION AND REMEDIES. — The points to be especially attended to are :—

1st. Any measures tending to lessen the quantity of eggs laid.

2ndly. Methods of cultivation which will destroy the egg or the grub in infested ground.

3rdly. Such applications of manure as may push on vigorous growth from the first sprouting of the seed, and also such special application in case of attack as may act rapidly—that is, be carried down at once to the roots, and thus invigorate the growth and carry it through the season when it is suffering from part of the supplies being cut off. Something may be added regarding means adapted to destroy the grub, but it appears that little confidence can be placed in any chemical applications that have been tried at present for this purpose.

1st. Measures tending to lessen quantity of eggs laid.—The parent fly frequents damp meadows, neglected herbage, and shady spots, such as are to be found by hedges and strips left at the sides of cultivated fields; and also under the shade of trees in open fields.

Rough-mowing the neglected ground, and burning the mixed grass and tops of weeds, would destroy a deal of shelter; and, in meadows, penning sheep on spots where the grass was not thoroughly eaten short would be a sure preservative. The biting of the sheep would clear off the shelter for the fly, and the trampling and general state of the surface consequent on their droppings would stop all oviposition.

Bush-harrowing meadows is a good plan to diminish fly-attack, as is also a good top dressing applied early in autumn.

Where the land is so damp that the Daddy Long-legs are present in most years to a hurtful extent, the ground should be drained.

It may be noticed, in crossing a large space of grass-land in autumn, that where the ground is dry and the herbage short, there are not any of these flies to be seen; whilst on the same land, with only the difference of shade or some cause for longer growth of herbage, they will rise in numbers before the passer-by. For this purpose—that is, of lessening the amount of flies—Rooks are of great service in the fields. Observation of a flock of Rooks on ground much infested showed they would catch and swallow the flies at a rate of four a minute, and in the smaller consideration of garden-ground, if a few tame and strong birds that can defend themselves from cats, like the Rook or Sea Gull, are kept about the gardens, they will do much good, as they are always on the search in the neglected corners which are the head-quarters of protecting weeds, and consequently of insect-vermin.

2ndly. Measures of cultivation which will destroy the egg or the grub in infested ground.—The grub is to be found most plentifully after a wet autumn, because this state of weather suits the flies, and in such circumstances the green crops are also suitable for oviposition, as well as the meadows and grassy spots which are most frequented in ordinary years; and also because the continued wet does not allow of the land being thoroughly cleaned.

Deep ploughing of leas or old pasture is a good practice, so as to bury the eggs and grubs deep in the soil. The absence of heat and air in the first case prevents or retards the hatching of the eggs; and in the second, some of the grubs are killed, and many are injured, by the unnatural surroundings and want of food, and thus made less hurtful to the crops.

A dressing of gas-lime on the pastures before breaking them up would prove serviceable. If it was applied in a fresh state the noisome smell would deter the fly from laying eggs, and the poisonous matter it contains would destroy the eggs already deposited, or the larvæ present in the soil. There would be ample time before seed-sowing for the necessary changes occurring to turn the deleterious substance into a safe manure, which changes can be effected in the soil, as well as when exposed to the air.

Paring and burning the surface with the grubs contained in it has been much resorted to, but the practice is open to some objections. It usually requires extra labourers, and it causes delay in cultivation, especially inconvenient in the North, where the climate allows little spare time in getting the leas ready for seed. Also the burning of the roots and plants tends to impoverish the soil, by the removal of organic matter, which would have benefited the crop much in its decay, and leaves instead nothing but ash.

Thorough cultivation such as would break up, or bury, or tear to pieces, all the clods of earth, tufts of grass, and other rubbish that the grub shelters in, or feeds on till the new crop is ready for it, is important, joined to such manuring as may be best suited on each different kind of soil to press forward the plant-growth. Attack in the early stage of growth, at the time when the food-stores in the seed are just used up, and the plant is beginning to depend on the first rootlets, are particularly to be guarded against, as a check sustained at this *first period* of growth is likely never to be entirely got over. Therefore such treatment as will cause a rapid healthy growth after sprouting will be found a good preservative, rather than a state of soil which will allow only of the plant slowly struggling forward, exposed for a much longer time in its weakest state to attack.

It is on cold wet land in general that this grub is most destructive, and it has been observed that it is upon the damp and clayey parts of a field that its attack is worst. The seed germinates more slowly on the colder soil, and the same amount of cultivation which is enough to put the rest of the field in good tilth is not sufficient to prepare an equally good seed-bed on this. The plant takes longer to develop, the growth is slower, and consequently, as it is not able by its growth to counter-balance the effect of the damage from the grub, it starves, sickens, and dies.

A well-drained soil, thoroughly worked and pulverized ground, and plenty of manure, are the things most likely to keep off attack; and with a special view to lay a foundation for strong growth where an attack is feared, it is recommended to use guano as a top dressing, along with the seed. This more particularly applies to clay-land, as the ammonia contained in guano is especially wanted on these soils.

3rdly. Applications that are of use when the grub is 'destructive to the growing crop. — With regard to *mechanical* applications. The grubs pierce the soil by means of their pointed heads, and draw themselves through by their rings; and therefore, though it is but an imperfect cure, something may be done by com-pressing the ground so that they cannot easily stray about under the surface, and waste and weaken even more than they totally destroy.

This may be done with a Crosskill roller, and rolling at night is also to be recommended, as many of the grubs are then on the surface. On a piece of ground that is so much infested that as many as one hundred and fifty or two hundred grubs are to be found in a square yard, not one will be seen during the day; but if the ground is examined after sunset, or before sunrise, the grubs will be found in numbers on the surface.

Where the land has been previously rolled so that no shelter is afforded to the grub, a good many may thus be crushed. Also they may be collected by a top-dressing of Rape-cake, and the roller passed over the ground early in the morning, with good results.

Hand-hoeing and horse-hoeing are noted as remedies,

the former being preferable, as it disturbs the working of the grub, and kills some, whilst it exposes others to the birds; at the same time this is "a dear remedy for a bad attack," as the grubs lie so close in amongst the roots of the plants. In a bad attack on a twenty-acre field of Peas, the outlay was at the rate of 4s. to 5s. 6d. per acre for five hoeings.—(E. A. F.)

The most successful application to the growing crop appears to be a top dressing of guano, or of guano and salt.

This latter applied at the rate of four hundredweight per acre, in the case of a very bad attack of grub on Oats, after Clover, was found perfectly successful in checking attack and running on a good growth, resulting in a fine crop, when all other means to stop the ravage had failed.—(S. F.)

When the *Cabbage* tribe are planted in ground infested by the grub, they should have their roots and stems up to the leaves dipped in a thickish mixture of clay, night-soil, and ammoniacal liquor from gas-works, or some other equally nauseous compound which the grub does not care to attack. As the stem of the plants just under the surface of the ground is usually the part attacked, it is a good plan, when the grub is observed to be active, to draw back the soil around the stem to a depth of a couple of inches, and put a ring of lime, soot, or even rotten manure, round the stem of each plant, which prevents attack and encourages growth.

The same traps that are used for the Wireworm are also effective in catching the grub of the Daddy Long-legs in garden cultivation.—(M. D.)

With regard to the action of *salt;* probably it does some good by driving the grub down; but, as an application by itself, it has failed in many cases, of which notes were taken during the bad attack of last year. As a dressing at the rate of seven hundredweight per acre to a grub-run field of twenty acres, it did no good.— (J. H. W.)

Special experiment by applying salt on three occasions

to a number of Cabbages, planted in pots and healthy, before the grub was introduced, showed that even with a quantity that killed the plants, the grubs were to be found at increased depths below the surface, these depths apparently regulated by the amount applied; but the grubs were in no way the worse.

It was found that the grubs might be immersed in strong brine for twenty-four hours without being killed. —(R. S.) From the above it appears that salt does not kill the grub, and in very wet weather, when the rain washes it down quickly, it is probably of little or no use; but in moderately dry weather it may be of some service in keeping the grubs from straying about on the surface, as it annoys them.

From experiments tried by Mr. A. Smetham, of Liverpool, by desire of Mr. S. Fitton, of Cheerbrook, Nantwich, which I am permitted to quote, it appears that a solution of four per cent. of carbolic acid and water killed the grubs in a space of from one to two minutes. In this solution, diluted to one-tenth of the above strength (that is, one part of carbolic acid to two hundred and fifty parts of water), the grubs lived about eight minutes. A few drops of the four per cent. solution of carbolic acid was poured on earth in which some of the grubs were contained, which immediately commenced wriggling about, and in a quarter of an hour were all dead. Practically, however, the application did not succeed in the field; the grubs were found to be presently within one inch of the surface on ground where the strong solution had been applied.

A solution of sulphate of copper (blue vitriol) killed the grubs in about eight to twelve minutes. Sulphate of iron (green vitriol) was more rapid in its effects.

Caustic potash appeared to have little influence.

The above experiments are of much value by showing how little these remedies can be depended on, some of which have often been tried, and time, valuable for checking attack at the beginning, thereby lost. It will be observed that the application that caused the most rapid destruction of life experimentally, failed to have any decided effect on the grubs in the ground, even when applied at a strength which, without the greatest care in using, would be destructive to the crop.

Immersion of the grub in nitrate of soda I have found to be followed by an immediate and violent discharge from the intestines; and nitrate of soda, being a stimulating manure, would be likely to do good on strong land, both by encouraging the plant-growth, and disagreeing with the grub; but applications of mere chemical poisons become so much weakened in passing into the ground, that their direct effect on insect-life by contact or by absorption is not to be trusted to as a sure remedy.

With regard to effect of cold, I was permitted to have some specimens frozen by artificial means, at Kew Observatory, down to a temperature of $-10°$, that is, ten degrees below zero, or forty-two degrees of frost; and although most of the grubs died, yet it showed that the grub could exceptionally survive even this temperature, to all appearance quite uninjured.

The grubs have been noticed frozen until quite brittle, and yet when thawed they were perfectly active.

With regard to power of bearing immersion, I found by experiment that although the grubs appeared to be dead after remaining in water for about fifty-eight hours, yet that they recovered after being exposed to the air; the exact time at which life was destroyed after being replaced in water was difficult to tell, but the whole time they lived from the beginning of the experiment might be considered about four days and nights; it certainly did not exceed five days and nights—that is, one hundred and twenty hours, for the grubs then burst.

The above experiments point to the fact that frost will not rid us of these grubs, and also that flooding land to get rid of them is not likely to be of much use unless the water is mixed with sewage, or some similar ingredient injurious to the grub, and is flooded on to the land so completely and for so long a time that the grub has no chance of escape. Where a field can be treated in this way, the plan has been found to answer well close to sewage-works.

With regard to want of food, some of the grubs placed in a vessel with a little earth, but no plant-food, were found to be alive and perfectly healthy after about three weeks; how much longer is not mentioned.—(S. F.)

Under drought, however, as far as experiment shows, the grub rapidly fails.

To recapitulate the above notes, we find that in natural circumstances the Daddy Long-legs Flies, that is to say, the flies of the different species of *Tipula* commonly so called, frequent damp places; the eggs are laid in damp places; the grub feeds and shelters itself in the dampest spots it can find, and only comes up to the surface at night when it is secure from sun-heat; and it will bear any amount of wet, short of continuous thorough immersion, for many days.

The application of strong chemicals which kill it when brought in immediate contact, fail when in the weakened state in which they must necessarily be applied practically; but some of these, as salt, and nitrate of soda, which are also serviceable as manures, have a deterrent effect. The grub does not like them, and to a certain extent keeps away if it can.

The remedies turn on meeting the above points by means varying with the soil.

1st. Drainage, and clearing away all unnecessary amount of wet neglected herbage; and such treatment of rubbish in the fields as may prevent it serving for food or sheltering places. 2ndly. Giving the plant a good start, and keeping up a healthy growth by ordinary measures of good cultivation and manuring; and 3rdly, in case of attack, although mechanical measures, as rolling, hoeing, &c., are of some service, mainly depending on such fertilizing application as will be available at once to the plant, and keep up its strength by the extra supply of food.

Ribbon-footed Corn Fly. Gout. } *Chlorops tæniopus*, Curtis.

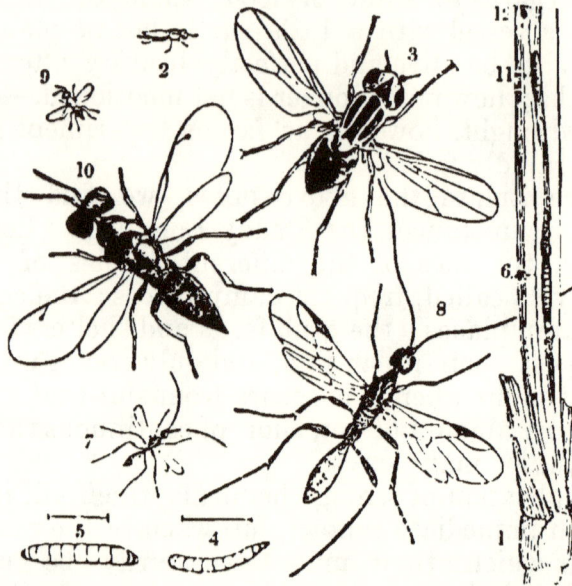

2—6, Larva, pupa, and fly of *Chlorops tæniopus*, nat. size and magnified parasite flies; 7 and 8, *Cœlinius niger*; 2 and 10, *Pteromalus micans*, nat. size and magnified; 1, 11, and 12, infested Corn-stem.

This Fly attacks various kinds of Corn, but in our own country appears to be most prevalent on Barley. In the summer of 1841 the Barley in Lancashire was destroyed to a great extent by these flies, and in 1846 it appears that from half to two-thirds of the crop was destroyed at localities named in the north of Lincolnshire, in Norfolk, Essex, and in Middlesex.

Where Barley is attacked by this insect the plants will be found stunted in growth, and late in ripening; with the stems sometimes distorted and swollen in the joints, whence the name of "Gout."

The ears, even at full growth, are generally still-sheathed, or partly sheathed, in the leaves, and on opening them the stem will be found to have a long pitchy brown furrow from the base of the ear down to the first knot in the stem. The ears are sometimes wholly abortive, sometimes with many grains absent, small, or mis-shapen; this being especially the case on the side of the ear on which the stem is furrowed.

The injury is caused by the egg of the *Chlorops* being laid either on the lowest part of the ear, or at its base, whilst the plant is still young; and by the feeding of the maggot hatched from it the growth is checked, and consequently the proper development of the ear is prevented.

The maggot is yellowish white, and legless, tapering to the head and blunt at the tail; it changes to a pupa in the sheathing leaves, and the very small oval rusty ochre-coloured fly-case may often be found lying in the black furrow caused by the gnawing of the grub. From this case the *Chlorops*, or Gout Fly, comes out towards the end of summer, as a small two-winged fly, about the eighth of an inch long, thick and stumpy in shape, yellow, with three black stripes along the back between the wings, and the abdomen of a greenish black, with black cross-bands. The wings when at rest extend beyond the tip of the abdomen.

These small flies may sometimes be found dispersed in stacks, or collected together in great numbers in some part of the stacks where corn may lie together from a badly-infested part of the field.—('Farm Insects,' 'Gard. Chron. and Ag. Gazette,' and ED.)

PREVENTION AND REMEDIES.—When a crop is seen to be much attacked, it is desirable to draw the injured plants by hand. This may be easily done, as their stunted or distorted growth points them out plainly, and they are chiefly to be found along the water-furrows or in the wetter parts of the field. By this means the risk ·of future attack is much lessened, but measures of prevention beforehand would be better. The injury is in consequence of the feeding of the maggot lessening the supply of food to the ear, a distorted growth being formed, and the date of development retarded; therefore whatever manure or treatment of the ground will promote healthy growth will be of service. One means towards this is thorough drainage of the soil, as it is upon the wettest parts of the field (by the water furrows), and on the most retentive soils (stiff clays), that the greatest amount of damage is noted as being done. Drainage would also make the soil warmer (by enabling it to retain the heat otherwise given off by evaporation of the surplus water); the application of manure would be more effective, and consequently the healthier plant-growth promoted would give quicker maturity, with increased quantity and quality of produce.

Barley, being a shallow-rooted plant, feeds in the surface soil, and comes quickly to maturity, therefore it needs that its food should be abundant and of a soluble kind; and it is for this reason that the application of any nitrogenous or ammoniacal manure, combined with phosphates, increases the yield to such an extent. This necessity is of course greatly increased when the stems are being injured by the maggot-attack, therefore the following applications are of use, in these proportions.

One hundredweight and a half of guano mixed with two hundredweight of superphosphate, applied at the time of sowing, when fears are entertained of attack, or one hundredweight nitrate of soda mixed with two hundredweight of common salt (chloride of sodium), applied when the braird has come well up, or when the crop is attacked, would prove beneficial. Salt would check any tendency of the plant to run too much to leafage, as a consequence of the action of the nitrate, and would stiffen the straw by promoting the assimilation of silica. The application of salt at the rate of from three to four hundredweight per acre has been recommended, but probably it is only of service upon soils in which organic matter is abundant, and where in a moist condition it principally acts as an antiseptic; thus checking too hasty decomposition, so that the plant-food is liberated in a slow but equal manner.

The application of lime in a caustic state would also be serviceable.

Wheat Midge.
Red Maggot. } *Cecidomyia tritici*, Kirby.

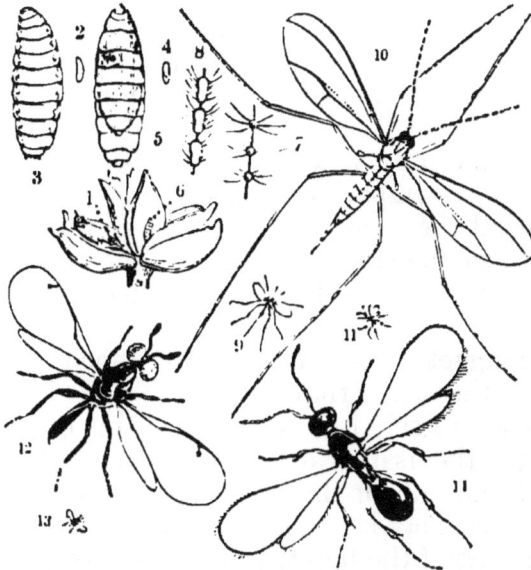

1, Infested floret; 2—6, larva, and cased-larva (? pupa). nat. size and magnified; 7 and 8, joints of antennæ, magnified; 9 and 10, *C. tritici*, nat. size and magnified. Parasite flies.—11 and 14, *Platygaster tipulæ ;* 12 and 13, *Macroglenes penetrans*, nat. size and magnified.

The grub of the Wheat Midge, known as the "Red Maggot," often does some harm, and at times causes a serious amount of damage to Wheat-crops, by injuring the young grains when forming in the ear, so that many of them never reach maturity.

In June, when the Wheat is in blossom, the female Midges may be seen laying their eggs, and are noted as being busiest at the work about eight o'clock in the evening. In the morning they may be found resting on the lower part of the culm of Wheat, with their heads downward, flying about, however, in great numbers near the ground when the stems are shaken.

The females are furnished with a long ovipositor, as thin as a hair, which they can extend at pleasure, and thus insert their eggs inside the florets. The eggs are oblong and transparent, and (with the help of a glass) may be found in little patches of from one up to twenty in number.

The maggots have been found ten days after the deposit of eggs was observed, some with their heads in the woolly top of the germ of the future grain, some inside the sheaths of the flower. It was not, however, observable that any meddled with the anthers of the stamens to an important degree, and it is therefore supposed that the injury to the grains in the ear is from the maggot so gnawing and drawing the moisture from the germ that the pollen cannot act sufficiently to fertilize it.

These maggots are lemon or orange-colour, more pointed at the head than the tail, and legless, but wrinkled transversely into folds, by means of which they can wriggle themselves along at pleasure. They are very small, only about the twelfth of an inch long, but their presence may be known by the altered colour of the lowest part of the floret, where they are to be found.

When full-grown, it appears that some of the maggots go down into the ground, but such as have not left the ears when the corn is cut are housed with it. This is important as regards treatment, but does not show any variation in the habits of the grub, as, if the corn had been left alone, it would have fallen, and the grubs thus come to the ground in the natural course of things. After a time they may be found in filmy transparent cases (see fig.), and turn to reddish pupæ, lighter at the end of the tail, from which the Midges appear in June.

The "Midge" is orange-yellow, or ochreous, with black eyes, and (as observable in the magnified figure) the longest vein of the wings is not forked.

The above species, that is, the *Cecidomyia tritici*, is the kind usually meant when speaking of the "Wheat Midge," but there is another kind that much resembles

it. This is the *Lasiopteryx obfuscata* (as identified by
the late And. Murray), differing from the above in the
fly being black, and the long vein of the wing being
forked at the end. The larvæ or maggots, and also the
pupæ of the two kinds, were indistinguishable; and, as
far as is known, their habits are also alike, and during a
series of observations made by myself on the habits of
the "Wheat Midge" in West Gloucestershire, in 1869
and 1870, it was found by Mr. Murray that all the
specimens sent to him, whether from chaff-heaps or
caught in the Wheat-fields of the neighbourhood, were
the *black*, not the *orange* Wheat Midge.—(See 'Gard.
Chron. and Ag. Gazette,' 1870 ; 'Farm Insects,' &c.)

PREVENTION AND REMEDIES. — The best method of
prevention consists of such cultivation and thorough
treatment of the surface of any field (known to have
been infested), after the crop is removed, as may destroy
these grubs or " Red Maggots" in the ground.

In Canada, where this maggot is much more hurtful
than with us, it is considered a complete cure to turn
down the surface of the field with the Michigan plough,
which, with the first turn-furrow, takes off about two
inches of the surface, together with the weeds and
stubble, and the insect-vermin in the roots, and deposits
them at the bottom of the furrow; whilst the second
turn-furrow raises another land-slice, and, depositing it
over the previous one, buries it several inches deep.
If the course of agriculture allows this to be left
untouched till after the usual time of appearance of the
Wheat Midge in the following year, it is found to
completely destroy the maggot.

In our own case, such ploughing and working of the
surface might be effected by having a skim-coulter
attached to the plough, constructed in such a manner
as would cut and lay an inch or two of the surface in the
bottom of the preceding furrow, thereby burying the Red
Maggots to such a depth as would render their coming
to the surface (or being brought to it by after cultivation)

G

very improbable; or broad-sharing might be adopted, which would tend to break up the surface soil.

The weeds and roots of all plants being either carted off or burnt, would help to destroy many of the maggots, and a large proportion of the remainder would be left on the top in reach of birds. These operations would be of service by putting the maggot in unnatural circumstances, which a large number of experiments have shown it to be particularly susceptible of injury from.

Another point of great importance is in regard to the grubs housed in the corn. Enormous quantities of these may be found in the chaff or the dust after threshing; and on neglected farms or small holdings where the chaff is often thrown in heaps, to decay in out-of-the-way corners, this treatment suits the Wheat Midges most admirably. The following June brings them out in clouds from the heaps to infest every Wheat-field near, and this practice, therefore, of spreading them is most objectionable.

It is also bad in another way by greatly increasing the opportunities of multiplication of the Midge.

It has been noted by careful observation that the Wheat Midges seen in the Wheat-fields were all females, consequently it may be supposed that few, if any, males were there; but from my own observations of the Midges over the chaff-heaps, I am led to think that pairing takes place immediately on the hatching of these Midges from the pupæ, and thus these vast collections are doubly hurtful. The heaps preserve the grubs through the winter, and from the quantity (probably) of both sexes that are hatched in one locality, the female is sent out in a condition at once to lay fertile eggs.

Where chaff is thrown away, it would be no greater loss to destroy it thoroughly by burning with other rubbish, or, if used as litter, it could be placed in the bottom of the yard, where the trampling amongst the wet droppings of the cattle would kill the maggots. Or again, the chaff might be placed in the bottom of the dung-pit, where the maggots would be effectually destroyed, and in this way there would be an increase of good farm-manure, with a saving of much more mischief from "Red Maggot" next season than is generally supposed. Simply burying the chaff in the ground is not at all a certain means of prevention.

Firing the stubbles (as mentioned under "Corn Saw-fly") is an effectual way of getting rid of many of the "Red Maggots" which may be found after the corn is cut, clinging about the lower part of the stalks.

In the American and Canadian experiments, it has been found of great service so to time the Wheat-sowing that the period of flowering should be clear of that of the appearance of the Wheat Midge. In the uncertain climate of our country any experiments depending on weather are of doubtful use, still the point is worth consideration. In this case the plant and the insect-development are not necessarily similarly affected by weather; a few days or hours warmth will bring the "Midges" from their pupa-cases, whilst the retarded plant takes much longer time to make up its lost ground; and the following observations, made in the wet summer of 1879, show instances of escape of the later Wheats from attack, because time of oviposition of the Midges was gone by before the ears were ready to receive the eggs.

On June 27th the Wheat Midges were abundant near Maldon, in Essex, whilst there were as yet no Wheat-ears on which they could oviposit, and the crop was not damaged.—(E. A. F.) In the part of Hertfordshire observed the "Red Maggot" was abundant only in the earlier fields (B. B.); and in the east of Norfolk, in the case of two small patches close alongside of each other, one that was autumn-sown suffered, but the other, which was spring-sown, was uninjured.—(A. S. O.)

Destroying such wild grasses as the Wheat Midge is known to frequent, especially the Wild Oat, or *Avena fatua*, is very desirable, and also the encouragement of the small insect-feeding birds.—(' Gard. Chron. and Ag. Gazette,' ' Report on Inj. Insects,' 1879, &c.)

Corn Sawfly. *Cephus pygmæus*, Curtis.

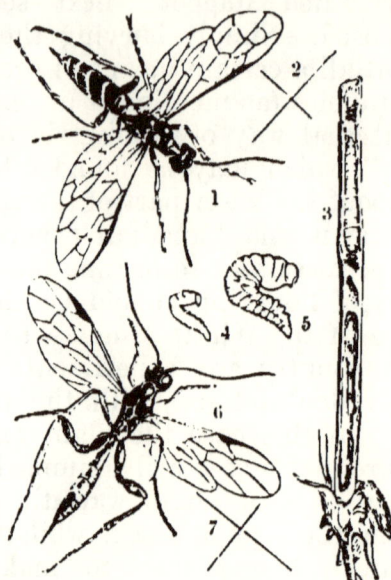

1 and 2. Sawfly, magnified and nat. size; 3, stem containing larva; 4 and 5, larva, nat. size and magnified; 6 and 7, parasite fly, *Pachymerus calcitrator*, magnified and nat. size.

This Fly attacks various kinds of Corn by piercing a hole in the stem whilst it is still young and soft, and laying an egg just below the first knot.

The young grubs feed on the inside of the stalk, and as soon as they are strong enough they gnaw through the knot and make their way upwards, sometimes through all the knots; they then come down again (always inside the stem), and about harvest-time, or a little before it, they cut the stem nearly through all round at the level of the ground.

These infested stalks may be known by the thin white ears standing upright and empty, or with few perfect grains; whilst the healthy plants are still green, and in some degree bending with the weight of the head.

When the maggot has thus travelled down the stalk

and nearly cut it through (so that nothing may prevent
its escape presently as a fly), it goes down into the
lowest part and spins itself a silken case, in which it
passes the winter. It changes to a pupa in the early
summer of the following year, and comes out as a fly
just in time to attack the new crops.

The maggot is about half an inch long, whitish and
fleshy, and has a horny brown head, with strong jaws.
It is very peculiar, for, though the larva of a Sawfly, it
has only the rudiments of three pairs of feet. It has
also a tubular appendage at the tip of the tail, which can
be drawn in and out like a telescope, and helps it in its
progress along the inside of the stalk.

The fly is four-winged, black, and more or less marked
or spotted with yellow on the head, abdomen, and legs.
The yellow is bright in the male; more of a sulphur or
ochre-colour in the female; and the wings are iridescent
in the male; more smoky in the female. —('Farm
Insects.')

PREVENTION AND REMEDIES.—The grub winters inside
the lowest part of the stalks of the stubble, therefore any
means whereby the stubble may be destroyed, or buried
too deeply for the fly to come up again after it has
developed, will be serviceable.

Where there has been a bad attack, it is desirable to
collect the rubbish, that is, the roots of stubble, and
either burn it in heaps on the field, spreading the ashes
afterwards over the surface, or cart it off the land with
the weeds, which, mixed with quick-lime or gas-lime,
forms a foundation for a good compost-heap, and is a
sure method of destruction of the grubs.

The clearing of the ground might be effected either by
a Biddle's scarifier or the broad-share paring-plough;
or a common plough, with its mould-board unattached,
and fitted with a share a foot in breadth, might be set
to work on the stubble after the grain-crop is carted, and
skim off the surface to the depth of three or four inches;
the harrows will shake the earth from the roots, and a

turn of the chain-harrows collect the refuse, which may be disposed of as above mentioned.

Burning the stubbles has been recommended, and is an excellent method of destroying the maggot; in some cases, by the elevation of the finger bar of the reaper, the stubble may be cut at such a height that the simple act of setting the residue on fire will burn the whole surface of the field, and prove a thorough remedy.

Commonly, however, the autumn-cleaning of the stubbles ought to be a sufficient remedy.

The Sawfly does much harm on the Continent, and is common in this country, in Corn-fields and on grasses growing in the woods in June; but is not known as one of our bad crop-pests.

Corn Thrips. (*Thrips cerealium*, Haliday.
(,,   *physapus*, Kirby.

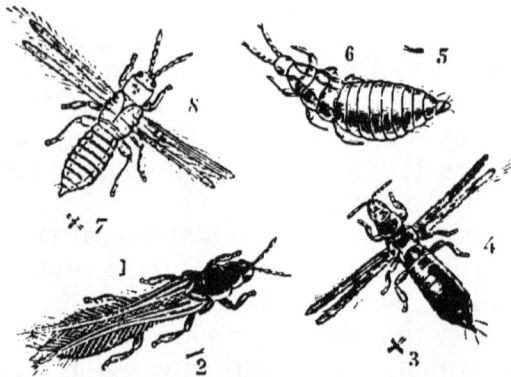

1—4, Corn Thrips (female). nat. size and magnified. (5—8, Potato Thrips. *T. minutissima*, nat. size and magnified.)

These very small insects are of the same kind as the little black creatures (hardly large enough to distinguish as insects) that are often exceedingly annoying by settling on the face in warm weather, causing a deal of irritation by running to and fro, and twisting in all directions. ·

If looked at under a glass, the perfect insect will be found to be blackish, hardly the twelfth of an inch long, the abdomen long, narrow, and smooth, with the tip bristly; the male without wings. The female has two pairs of long narrow wings, fringed with long hairs, and curving outwards, so that when they are laid straight along the body at rest the tips are apart. The feet are very short and stumpy, without claws.

The two earlier stages of grub and pupa much resemble the perfect insect, excepting that the grub is deep yellow, and has no wings; the pupa is of a paler yellow, with whitish cases for the wings which are not yet developed.

The Thrips are to be found on Wheat-ears in every stage of growth, from June till the ripening of the ears, also sometimes in the sheathing-leaves of the stem.

The common position, when feeding, appears to be with the sucker fixed in the furrow, and towards the lower part of the grain, thus drawing away the juice and causing the seed to shrivel. They are said also to do harm by sucking the juice from the stalk above the knots whilst it is still soft and tender.—('Farm Insects.')

PREVENTION AND REMEDIES.—I find very little bearing on these points, but it is stated that Thrips do most mischief to late-sown Wheat, the early-sown crop being too hard at the time the Thrips appear for them to injure it. Also these insects are to be found near marshes, and in the damp hot weather accompanying thunder-storms they have been noticed as especially injuring the crop in the part of a field to the north of a high hedge, and also the Potato Thrips, figured above, have been noted as most abundant on plants sheltered from the wind.

From these observations it would seem that a well-drained soil, properly cultivated and free from weeds, with surrounding hedges well trimmed down, so that the sun and wind might have free play, would be the best preventive, by inducing an early ripening of the corn that would be beyond the power of the insect to injure, and would also do away with the damp close spots that, in some cases at least, it had been found to frequent.

It does not appear that any of the applications, such as dusting with sulphur, the use of tobacco-water, &c., which are serviceable in the garden, can be remuneratively used in the field.

Wireworms. Grubs of Click Beetles.	*Elater lineatus*, Linn.
	,, *obscurus*, ,,
	,, *sputator*, ,,
	,, *ruficaudis*, Gyll.

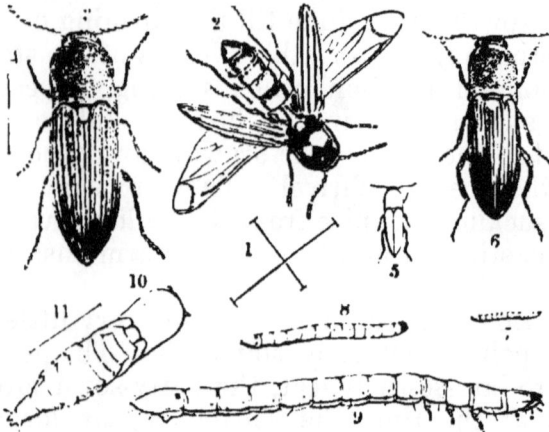

1 and 2, *E. lineatus*; 3 and 4, *E. obscurus*; 5 and 6, *E. sputator*, nat. size and magnified; 7, larva of *E. sputator?*; 8 and 9, larva of *E. lineatus*, nat. size and magnified; 10, pupa of Wireworm (lines show nat. length).

Wireworms may perhaps be said to do the greatest amount of mischief of any of our farm pests: they destroy root, grain, and fodder-crops. From their method of gnawing the roots or underground shoots, and then going on to another plant, they waste and destroy far more than they need for food; and as they live for several years as grubs, and feed during these years on almost every kind of crop that is commonly grown, their ravages are of a very serious kind.

The Wireworm is the grub of the long narrow greyish brown, or blackish beetle (see figs. 1—6), often seen during summer in grass-fields, commonly known as the Skip-Jack, or Click Beetle, from its power of regaining its position, when laid on its back, by a spring or skip, accompanied by a sharp click.

Figs. 7 and 8 give the size of two kinds of these grubs which are called Wireworms, from their likeness in toughness and shape to a piece of wire. Like it they are very smooth and shining, and somewhat cylindrical; but a little flattened, so that (like a wire that has been pressed by a weight) they have a blunt edge at each side. The colour is ochreous-yellow, turning to a darker tint after death.

The Wireworms have three pairs of short legs, one pair of these being placed on each of the rings imme-diately behind the head, and they have also a sucker-foot below the tail.

The egg from which this grub is hatched is laid either in the earth close to the root of a plant, or between the sheathing-leaves near the base of the stem. On being hatched, the grub or "Wireworm" eats into the stem just above the true root, about an inch below the surface of the ground, and sometimes eats its way up the middle of the stalk, even above the surface of the earth.

The Wireworms are said to live five years in the grub state, but the length of time probably depends on the supply of food. Where they are well fed, it is supposed that they only take about three years before changing to the pupa. But however this may be, with the exception of any temporary pause in winter (when they go down deeper and deeper into the ground as the frost increases), they feed voraciously near the surface till the time has come to turn into the chrysalis (or pupa). Then they go deep into the soil, and form an earth-cell in which they change, and from which the perfect Beetle comes up through the earth in two or three weeks, probably appearing about the first weeks of August; or they may pass the winter in this state, and the beetles develop from the chrysalis in the following spring.

Of the many kinds that are to be found of these beetles, only four are noticed as being particularly hurtful to crops. These were formerly all known scientifically as different species of *Elater*. *E. sputator*, fig. 5 (and 6, magnified), is the smallest. It has the head, and part of the body behind it, black; legs rufous; and the wing-cases dusky.

E. obscurus (fig. 3, magnified) is larger and pitchy, covered with ochreous down or hairs; so that perfect specimens appear dull brown all over, and rubbed ones blackish. It has black thighs, and the shanks and feet rusty.

E. lineatus (fig. 2, magnified) is very like the preceding, but greyer; it has the wing-cases striped, and the legs rusty red. The Wireworm of this species is often found in dung and vegetable earth.

These three kinds are now generally known as *Agriotes*, instead of *Elater*.

The fourth species, now known as *Athöus*, has its special name of *ruficaudis*, from the red colour of the abdomen and tail; this is larger than the others.

These four kinds of Click Beetle are, however, alike in all important points in their manner of life.

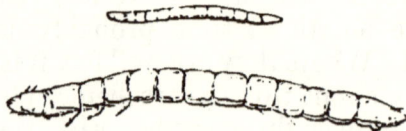

Elater (Agriotes) obscurus, nat. size and magnified.

The reader is particularly requested to notice the figure and description of these grubs or larvæ, that is, of the "*true Wireworm*," that he may distinguish it from the grubs of other insects which are not altogether unlike it, and from insect allies which pass under the name of *false Wireworms*.

The Wireworm has *six* true legs; this distinguishes it from the grubs of the Daddy Long-legs, or Crane Fly (Leather Jackets), which have *none*. Also the Wireworm has *only* six legs (besides the sucker-foot at the end of the tail); this distinguishes it from the Centipedes and Millipedes, which have many.

Of these the Centipedes are chiefly animal-feeders, or feed like worms on the earth; the Millipedes, of which four kinds are figured below, feed on plant-roots, on decayed vegetation, and more or less on animal matter, such as grubs, earth worms, &c. — ('Farm Insects'; 'Skip Jack, and the Wireworm,' by A. M., &c.)

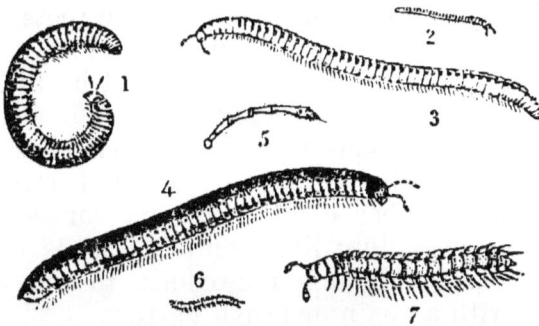

FALSE WIREWORMS.

Snake Millipedes.—1, *Julus Londinensis;* 2 and 3, *J. guttatus,* nat. size and magnified; 4, *J. terrestris;* 5, horn; 6 and 7, flattened Millipede, *Polydesmus complanatus.*

PREVENTION AND REMEDIES. — Soil from broken-up pasture-land or Clover-leys often swarms with these pernicious grubs, and it is from this infested ground that the most serious damage arises both to garden and field-crops.

With regard to garden-crops, the broken turf should never be used without having been thoroughly examined, so as to make sure there are no Wireworms in it. If it is only wanted in small quantities, this can be easily done; the turf can be broken by hand into small bits, so that its state can be seen.

Where a large quantity is wanted, it is a good practice to throw the turf in a heap after being mixed with fresh gas-lime, a dressing of gas-lime being also spread over the heap.

The sulphide which is present in *fresh* gas-lime will kill all the vegetation and the insects, and by the time the turf is required for use it will be converted by the action of the air and moisture into sulphate of lime (gypsum), which is serviceable in various ways as a manure.

If turf-heaps are allowed to get covered with grass, it is of no use, so far as getting rid of Wireworm is concerned, to have them at all; for these grubs will be as perfectly well suited there as in the field. If we are to get rid of the Wireworm feeding at the roots, we must get rid of the growing grass or plants, and any treatment is good that brings this about, whether it be turning the heaps, liming, burning, or otherwise.

In cultivation of fields broken up from grass-land or clover-leys, something may be done beforehand to diminish the number of insects, and consequently of eggs laid, by spreading lime-compost, ashes (on heavy land), or any dressing obnoxious to insects, and harrowing with a chain or brush-harrow.

Paring and burning is a good practice, so far as getting rid of the grubs is concerned, as great numbers of the grubs and other insect-vermin are thus destroyed, and also a large amount of the live grass-roots and weeds that might feed such grubs as remained in the ground. The rubbish should be burnt as soon as possible, or the Wireworms will quickly leave it, go down into the earth, and escape. The habit of the Wireworm to go down deep—even as much as a foot below the ground in winter—should also be remembered, as a paring that would clear off a large proportion of the Wireworms in warm weather would very likely pass above almost all of them in winter, and leave them uninjured to come up to the surface in spring.

A summer fallow, and burning all rubbish on the ground is also recommended. By this means the grubs at the shoots are destroyed at once, and those in the ground die for want of food. The application of gas-lime is also useful here; it may be mixed with three or four times its own bulk of earth, and applied as a dressing; or it may be spread thinly over the ground, and well mixed with it a short time before the crop is put in.

It is probable that, where the land can be left free for a sufficient time afterwards, it would answer well to put a strong dressing of gas-lime fresh from the works on the ground after ploughing and cleaning. This would destroy everything it touched, and by spring would be changed (as noticed above) into a good manure.

But, whether the object is carried out by special applications, or by the common operations of thorough cleaning of the broken-up ground and burning the rubbish, it is of great importance to clear the field thoroughly of the Wireworms before the new crop is put in. No time is gained, *for us*, by putting it in, *to them;* they will take each crop in succession, and waste or destroy till they begin their change to the condition of Click Beetles.

Strips of grass or clover should not be left growing in the field, or at the sides, for the Wireworms feeding in these will spread themselves round and damage the neighbouring crop.

Such methods of cultivation as will ensure a strong healthy growth are of great importance; the Wireworms injure and weaken a great deal more than they destroy; and if the soil is so prepared as to ensure available food to the plants, they will push on and get over the attack, where weak plants would sink under it. This is especially the case with grain-crops.

Drilling manure with the seed will help in this matter. Probably any manure that acts rapidly will be of service, but Lawes's Turnip Manure has been found to answer well with Barley on a badly-infested piece after dead or bare fallow, the parts of the field not thus treated having more than half the plants destroyed.— (B. B.)

A mixture of guano with superphosphate of lime, drilled with the seed on pasture-land broken up the previous year, has similarly brought a good crop, whilst the rest of the plants on the field perished by Wire-worm: in this case Wireworm was found between the drills, which seems to point to the safety of the crop being from the dislike of the grubs to the manure, as well as to the increased strength of growth. Dissolved bones drilled with the seed also do good.

Soot and also guano have been found to stop the mischief in bad attack on Oats. The soot was applied at the rate of sixteen bushels per acre, the guano at the rate of two hundredweight, "all in a pouring rain." Many of the patches that were apparently destroyed put out new roots at about half an inch below the surface, and the crop was excellent.

In this case there would be benefit from the manure being washed down into the soil for immediate use, and if on the first signs of attack, before the strength of the plant is gone, any kind of liquid manure that may be preferred was applied *at once*,—by the liquid manure-cart, where this is practicable,—it would do much good, partly by driving some amount of the Wireworms away, but chiefly by giving the plant strong food in a form that it could take up at once, and so counterbalance the lessened supplies conveyed up the partly-eaten stem.

Any stimulating manure, whether chemical or other-wise, suitable to the soil, which is easily soluble by rain and which would thus come quickly into action, would be of service.

Nitrate of soda and common salt, mixed in the proportion of one hundredweight of the first with two hundredweight of the second, have proved useful for this purpose; but with regard to special applications there is much difference of opinion, probably arising from different circumstances as to soil, time of application, and amount used.

Soda-ash is said to be of use. Salt applied at the rate of twenty bushels per acre to badly-infested land did well. In this case the salted part bore a beautiful crop, whilst that to which salt was not applied did not bear a fourth. Any manure which dissolves quickly and that is suitable to the crop, and can be applied so as to act at once, will do good.

Rape-dust is a good stimulating manure, and Rape-cake has been found of use, applied in the proportion of five hundredweight to the acre, crushed into about half-inch lumps, and mixed well with the soil. This acts by drawing the Wireworms from the crop; the Wireworms are stated to collect in vast numbers in it, and the plants are thus freed.

With regard to *mechanical* applications, one of the common remedies used among corn-crops is, rolling with a heavy roller so as to solidify the surface, and thus prevent the grubs from travelling through the ground. The Cambridge or ribbed roller is serviceable for this, or Crosskell's clod-crusher, where it can be used on lighter land.

The remedies used amongst root-crops are, drill-hoeing, horse-hoeing twice in a place, hand-hoeing close to the rows, and chopping-out to stop the progress of the Wireworm along the drills; here the object of the treatment is by stirring the soil, to encourage the growth of the plants, and to harass and disturb the Wireworm. —(E. A. F.)

Treading the Barley with young sheep (tegs) is of use in keeping the Wireworm from "running."—(T. H.)

In attack of Wireworms on young beds of year-old seedlings of forest-trees, it has been found of use to scrape the earth back from the collars of the young plants to a distance of about six inches, so as to lay bare the larger roots, and hand-pick and destroy the grubs. The roots were then dusted freely with equal parts of lime and fresh dry soot, and fresh soil used to cover them; the old soil being removed and charred, to kill any Wireworm that might remain in it.—(M. D.)

In garden-cultivation, one exceedingly important means of prevention is to avoid the use of infested turf. Deep-trenching is a good means of getting rid of the Wireworm (P. L.); and where they are numerous and injurious in gardens, great quantities may be caught by burying Carrots or Potatoes two feet apart and six inches deep in the borders, and examining them and destroying the Wireworms found every other day (P. B.); or pieces of Carrot or Potato may be run through with a stick (to save trouble in search) and buried nearer the surface, and examined daily.

Paraffin used in the proportion of one part to fifteen of water, has been tried in the north amongst garden-crops, especially amongst Carrots, and has proved an effectual remedy.

Paraffin has also been found of service in stopping bad attack on Turnips. In this instance a quantity of dry sand was procured and moistened with paraffin just sufficiently for it not to clog, but run freely in the hand; this was strewed lightly along the centre of the drills, so as to fall directly above the roots of the plants, throughout a four-acre field, and the result was satisfactory. The attack was stopped, and a good crop was obtained. —(J. K.)

Two crops which Wireworm are considered by most observers to dislike are White Mustard and Woad, and in experiments on fields of fifty and forty acres respectively, of Wheat which suffered from Wireworm, the portions which had been under White Mustard were free from attack. If this should be commonly the case, a crop of White Mustard would be very desirable to clear the Wireworm from broken-up pasture.

The natural enemies of the Wireworm are the mole, which does good service in this matter to a considerable amount; rooks, plovers, peewits, or lapwings also assist in keeping them down.

HOP.*

Hop Aphis. Green Fly. { *Aphis humuli*, Schrank.
Phorodon humuli, Schrank.
„ var. *Malaheb*, Fonsc.

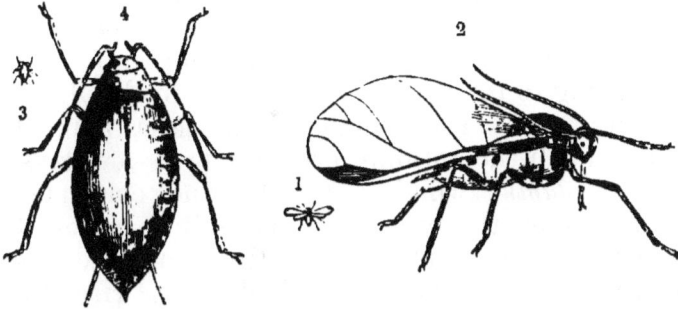

1 and 2, Female Aphis, nat. size and magnified; 3 and 4, larvæ or "nits," nat. size and magnified.

The Hop-plant has many insect-foes. Chief amongst these is the Green Fly, or Aphis (*Aphis humuli*), which in some seasons has caused the almost total destruction of the crop of Hops.

The general appearance of these pale green Aphides is too well known to require description, but it may be observed that the genus *Phorodon*, to which they belong, is distinguishable from others of the *Aphidinæ* by the horns being hardly longer than the body, together with the lowest joint being toothed or gibbous, and the tubercles on the forehead each having a strong tooth. The legs are short, and the honey-tubes long.

The variety "*Malaheb*" is distinguishable from the species *humuli* by the females bearing living young being larger, and, in the case of the wingless kind, of a yellower green than *humuli*; in the case of the winged kind, of a brighter green, with the head, horns, and legs

* In the papers regarding Hop-insects, I am much indebted for valuable information to C. Whitehead, Esq., F.L.S., of Barming House, near Maidstone.

H

(excepting the thighs) black, instead of brown or brown and green; and the markings on the abdomen dark, instead of black, as in the case of *humuli*.

The male of *P. humuli*, which is winged, may be known from the winged female by being smaller, but with large head, small abdomen, longish horns and wings, and the colour wholly pale green, save some olive markings on the back: this is to be found in the beginning of September on the Hop-plants.

According to the common rule of life in the *Aphidinæ*, the appearance of the male in the later part of the summer, or in the autumn, precedes by a few days that of the *egg-laying* wingless female, and concludes the generations of the year. It is only between these that pairing takes place, and the fecundated female subsequently lays the eggs which give rise to the countless winged swarms of the following year, all produced without the further intervention of the male, until the season again closes with the autumn appearance of the winged male and wingless oviparous female.
—('Mon. of Brit. Aphides,' &c.)

The young, or larvæ,—commonly known as "lice" or "nits,"—which may be seen whilst being produced alive as early as the middle of May, are at first of a semi-transparent greenish white, afterwards they are green, and much resemble their parents, excepting in not having wings.

The viviparous females always appear in greater or less numbers in the Hop-gardens about the second or third week in May. If the conditions are favourable, they then deposit their first brood of larvæ on the leaves of the plants, and, from the early date at which these young have the power of producing descendants, increase goes on at a rate which words do not convey, but which the state of the attacked plants shows only too plainly.

The result of this is, if the Aphides are undisturbed, what is known in Hop-districts as "*a blight*" occurs. The powers of the leaves are exhausted by the millions of insects drawing out the juice with their suckers, and

the pores of the leaves are choked by the fluid voided by
the Aphides, or by honey-dew; so that they cannot
perform their natural functions, and the growth of the
plants is consequently checked.

In some seasons the Aphides do not remain to breed.
It is supposed that they do not remain unless the plants
are in a peculiar condition—an unhealthy or abnormal
state in which the sap is grateful to their taste.—(C. W.)

PREVENTION AND REMEDIES.—"When it is seen that the
winged flies stick persistently to the under surface of the
leaves, and that larvæ or 'lice' have been produced, the
planters begin to wash the plants. This can hardly be
done too early, as when the lice swarm it is difficult to
dislodge them all. Washing is done by means of a
large garden-engine, fitted with a pump, and a long
length of gutta-percha hose on each side, having a single
jet, or rose, or spray syringe, which can be directed
under the leaves, and round the bines, thoroughly cleaning the plants. This engine is drawn along the rows or
alleys by three men, one of whom also pumps; the
others take a hose each, which is long enough to enable
them to wash three rows.

"The wash is usually composed of from fourteen to
twenty pounds of soft soap to one hundred gallons of
water; some add the juice of a quarter of a pound of
tobacco; others think that the tobacco kills the 'negur'
—the larva of the Lady-bird, or *Coccinella*, which feeds
on the Aphides.

"The cost of washing varies from thirty to thirty-five
shillings per acre each time it is done. Great care must
be taken to wash every leaf underneath, and the process
generally must be repeated twice, or even thrice.

"The great points are—to begin early, to have the
plants thoroughly washed, and to continue to wash
until the leaves are clear from lice. In ordinary seasons
washing is efficacious, if thoroughly carried out, and
there are many planters who have made good profits by
judicious washing. If, however, the weather is wet and

cold, as in 1879, washing seems to be of little avail."—
(C. W.)

The following recipe for Hop-wash, used by an extensive grower, has also been found reliable in dry
weather :—To thirty-six gallons of water in a copper add
sixty pounds of soft soap. Then add either fourteen
pounds of bitter aloes or two pounds of tobacco, and
boil together. For use, add thirty-six gallons of water
to every gallon of this liquid.—(J. W.)

In some cases a little soda is found serviceable, mixed
with the solution of soft soap.

Quassia has been experimented with at a strength of
four ounces to one gallon of water, but proved quite
useless, either to kill or to drive away the Green Fly.—
(T. H. H.)

Many methods of treatment, such as dipping infested
shoots, fumigation, &c., are suggested by various writers,
but appear of very doubtful use. At present, though
"remedies" are found, as above mentioned, to cure or
mitigate the attack, no "means of prevention" of it
have been discovered.

It is found that circumstances of locality, whether
from soil or surroundings, exercise an influence in the
amount of attack, as "In some districts the Hop-plants
are more liable to be blighted than others, and in most
districts there are 'lucky' farms upon which the Aphis-
blight or mould rarely affects the plants. A hedge or a
stream frequently forms a line of demarcation between
Hop-land that is liable to blight and that which ordinarily
escapes blight."—(C. W.)

It is also observed, "Flat low lands, although the best
land, and generally the most productive and sheltered
from wind, are more subject to be blighted by Aphis
than higher and more exposed situations."—(S. R.)

Weather influences have great effect on the Hop.
Under favourable circumstances it grows rapidly, especially in warm nights—as much as four inches of growth
having been recorded as taking place in one night on
bines. East winds in the spring are unfavourable to

the plants, and favourable to Green Fly. The sunshine, and warmth in sheltered positions which often accompany these winds, alternate with frosts at night that check the growth of the tender succulent shoots, and thus a state of sap is produced peculiarly suitable to the Aphides, which increase rapidly under these circumstances.

Practically it has been known as far back as Tusser's time, that—

> "The wind in the North, or else Northerly-East,
> Is bad for the Hop as a fray at a feast,"

and scientifically the recent observations of Mr. G. B. Buckton, as to the more rapid development of Aphides on plants of which the sap is sickly, and the amount of food to be drawn from them insufficient, points to the reason of the increased attack, but how this knowledge is to be utilised does not appear at present.

Amongst natural means of protection we have some help from the Aphis-eating birds, but the chief assistance is from the *Coccinellidæ*, the beetles well known as "Lady-birds," which feed (especially in the larval stage) voraciously on the Aphides.

Lady birds and chrysalids; 1—4, egg and larva, nat. size and magnified; 7, *Coccinella bipunctata*; 8, *C. dispar*; 9, *C. septempunctata*.

The above figure shows the appearance of the slaty grey six-legged grubs (magnified, with line giving length when full-grown). These grubs are prettily marked with scarlet and yellow, and when full-fed, which is in about

a fortnight or three weeks, they hang themselves up by
the tail, and turn to a shiny black pupa or chrysalis,
spotted down the back with orange, from which the
beetle (known as the "Lady-bird") comes out in about
another fortnight or three weeks. The figures give the
common red Lady-bird, distinguishable by its seven
black spots, and two smaller kinds.

These should be by all means protected, and especially
when they appear in the vast swarms in which they
frequently follow on a special outbreak of Aphides, and
in which to our great injury they are liable to be swept
up and destroyed, as in the instance of their great
appearance in 1869.

Hop Bug.   *Lygus umbellatarum.*

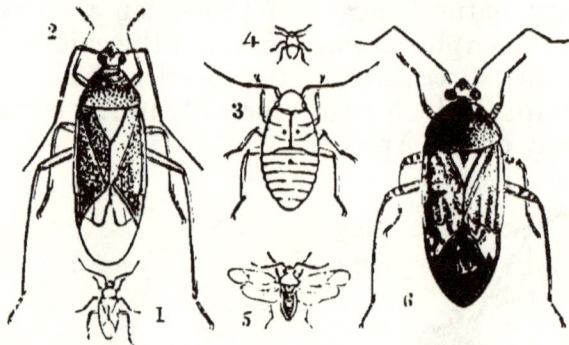

1 and 2, Potato-bug, *L. Solani;* 3 and 4, pupæ of do., nat. size and
magnified; 5 and 6, *L. umbellatarum,* nat. size and magnified.

In the case of the plant-bugs, whether of *Lygus
umbellatarum* or other species, the injury to the plant is
effected by the insect inserting the sharp beak-like
sucker into the shoot, leaf, bine, or whatever part it feeds
on, and drawing away the juices. Thus it not only
weakens the plant by the quantity of sap it takes away,
but it also injures the tissues and disturbs the regular flow
of sap by the immense quantities of little holes it drills
with its sucker into whatever part it may select for attack.

The *Lygus umbellatarum* (of Panzer, figured nat. size at 5, magnified at 6) is for the most part red or rosy on the head, horns and body between the wings ; a triangular space behind this portion white, with black at the base ; abdomen shining-black above, with ochre-coloured margin. The wing-cases are clouded with red, and the horny or membranous part near the tip has a smoky border. The wings (which when at rest are folded beneath the half-solid, half-transparent wing-cases) are large and iridescent, with dusky nerves ; legs varied with ochre and red ; shanks tipped with brown, feet pitchy.

The plant-bugs in their earlier stages are similar in shape to the parents, excepting in the absence of wings, and similar in their method of feeding ; but in the case of this *Lygus* it has hitherto only been noticed feeding in the pupa state on the Hop.—('Journal of Royal Ag. Soc.,' vol. x., p. 79.)

PREVENTION AND REMEDIES.—In the autumn, winter, and spring, Hemiptera (Plant-bugs) may be found about the roots of plants, in tufts of grass and in moss, among dead leaves and the *débris* of hay-ricks, and corn-stacks, and in field-rubbish.—(Brit. Hemiptera.)

This plant-bug, the *L. umbellatarum*, is to be found on Potatoes, and is said by Curtis to be found on umbellate plants later in the year ; and also on grass in May, and another nearly-allied kind is found on Nettles. As far as appears at present, the only way of diminishing the attack (excepting by the syringings and dustings used for other Hop-pests) is by clearing away the rubbish it shelters in during the winter, and the weeds that it feeds on in summer.

Hop Flea. $\left\{\begin{array}{l}\text{\textit{Haltica concinna}, Curtis.}\\\quad,,\quad\text{(syn. \textit{dentipes}).}\end{array}\right.$

This species of Hop Flea, known as the Brassy or Tooth-legged Turnip Flea, is very like the common

Turnip Flea or Fly-beetle, but differs in being more
oval, convex, and shiny. It is of a greenish black
colour, with a brassy or coppery tint; the horns are
only half as long as the body, and of a pitchy colour,
more rust-coloured towards the head; the wing-cases
have ten lines of deep dots along each; the legs are
black, but bright rust-colour at the base of the shanks,

1 and 2, Hop Flea, nat. size and magnified; 3, hind leg, magnified.

and the shanks on the second and hind pairs of legs are
toothed below the middle, whence the common name of
"tooth-legged" and the scientific synonym of *dentipes*.
This flea-beetle infests Hop-grounds and Turnip-fields,
and is also to be found in hedges, nettles, and grass.—
(' Farm Insects.')

It does not appear certain, however, that it is the *only*
kind of flea-beetle that infests our Hop-grounds, and the
following observations on the habits of the "Hop Flea"
are given with the note that (although resembling the
species described above by Curtis as *Haltica concinna*)
this may prove to be a kind especially confined to the
Hop :—

"These flea-beetles hybernate in the perfect state in
the ground close to the Hop-hills, or in the hollow dead
bines left on the stocks, or in the pieces lying on
the ground near them. They emerge in the early
spring, and attack the shoots of the Hop-plant, piercing
them as soon as they appear.

"If the weather is cold, and the shoots unable to grow away rapidly, the Flea occasionally causes serious injuries, and makes them stunted. They are especially active in dry seasons and when the land is rough. In wet seasons, and in growing seasons, when the shoots go quickly up the poles, they do not cause much harm in the early stages of the plant-growth. But later on in the summer, after a very dry season which has been favourable to their increase, they get into the cones and deposit their eggs. From these eggs the larvæ—little white maggots with six pectoral feet—are hatched in about ten days; they immediately begin to burrow in, and feed on the stalks or 'strigs,' causing their decay, and making the bracts of the cones lose colour and become disintegrated.

"During the spring and earlier part of summer the eggs are laid in the bine, or under the skin of the leaves; and one female (*vide* J. Curtis's 'Farm Insects') will lay about one egg daily, so that they are not extraordinarily prolific."—(C. W.)

PREVENTION AND REMEDIES.—"To check the Fleas, planters dust lime or soot over and around the Hop-hills when the shoots are low, but there is no remedy adopted against their onslaughts after the bines have been tied to the poles.

"One means of preventing the spread of these beetles is to have all the pieces of old bines carefully removed from the Hop-garden after Hop-picking and all the cuttings after the hills have been 'dressed,' and to move and pulverize the ground as early as possible in the spring. The Flea is very injurious to the bine in early spring, particularly in dry weather, and where the land is rough and badly cultivated."—(C. W.)

The following remedies have been suggested, namely, to cover up the young shoots with four to five inches of fine mould, which gives them security from being injured by the Flea for several days, when the bines will have acquired more strength and grow more rapidly away from attack; and where much injury is done, the application of a little rich manure, as Peruvian guano, is recommended.—('Farm Crops.')

In the 'Report on the Flea-beetle of the Vineyards of the United States of America,' where great losses are caused by the species *Graptodera* (*Haltica*) *chalybea*, Illiger, besides the above mentioned measures of clearing all pieces of old bine, broken wood, and rubbish, under which the beetle can harbour in winter, it is stated that when attack has commenced, good may be done "by syringing the vines with a solution of whale-oil (= fish-oil, Ed.), soft soap, two pounds of soap to sixteen gallons of water."

The following method, which was found to keep the flea-beetles in check in a large vineyard at Arlington, Virginia, might be serviceable:—

"A strip of cotton-cloth three feet by six, kept open by cross-sticks at the end, is thoroughly saturated with kerosine and held under the vine, which is shaken by the supports being struck. The beetles fall readily by the jar, and contact with the kerosene sooner or later destroys them. . . . . After striking the sheet the flea-beetles show no disposition either to fly or jump."—(Report of the Commissioner, Department of Ag., U. S. A., 1879.)

It is stated that, with these sheets, three boys rapidly clear the vines over a large space of ground, and in case of bad attack the method would be worth a trial in our Hop-grounds.

Hop Frog Fly. *Eupteryx picta*, Fab.

Little green and yellowish green insects, like tiny Grasshoppers, weaken the Hop-plants much in some seasons by boring holes in the leaves and bines, and sucking the juices. They are endued with mighty powers of jumping, and are called "Jumpers" in Kent. —(C. W.)

These are probably *Eupteryx picta*, a species very similar in form to *E. Solani*, figured amongst insects injurious to Potatoes, but larger, and more spotted. They are of a yellow colour, spotted with black on the head and on the body between the wings; the abdomen is black, with yellow margins to the segments. The upper wings are clouded with brown, and have the base, tip, and various spots, yellow; the under wings are transparent, with brown veins; the legs are sulphur-colour.

The pupæ resemble the perfect insects in shape, excepting that they are as yet wingless: and are of a buff colour, with dark eyes, and tips to the feet.— ('Farm Insects.')

PREVENTION AND REMEDIES.—When "Jumpers" are numerous, men are sent into the Hop-gardens with boards covered with tar, held at the leeward side of the bine. The poles are violently shaken or struck, and the Jumpers, taking tremendous "leaps in the dark," fall on the tarred boards held underneath.—(C. W.)

This Frog-fly also inhabits Burdock and Nettles, most objectionable weeds in a Hop-district, being frequented by various Hop-pests.

Hop-dog.
Caterpillar of
Pale Tussock Moth. } *Dasychira pudibunda*, Linn.

The caterpillar of this moth (commonly known as the Hop-dog) is often to be found in Hop-grounds.

It is of a delicate green colour, with a good-sized bunch of yellow hairs standing up like a brush on the back of the fifth, sixth, seventh, and eighth segments, each of which have a stripe of velvety black between them; a longer and more slender tuft of hairs at the extremity projects backwards like a rose-coloured tail.

The caterpillars feed during summer on the leaves

of various kinds of trees, also on the Hop, and about
September spin a slight web amongst the leaves in
which each turns to a hairy chrysalis, from which the
moth comes out in the following May.

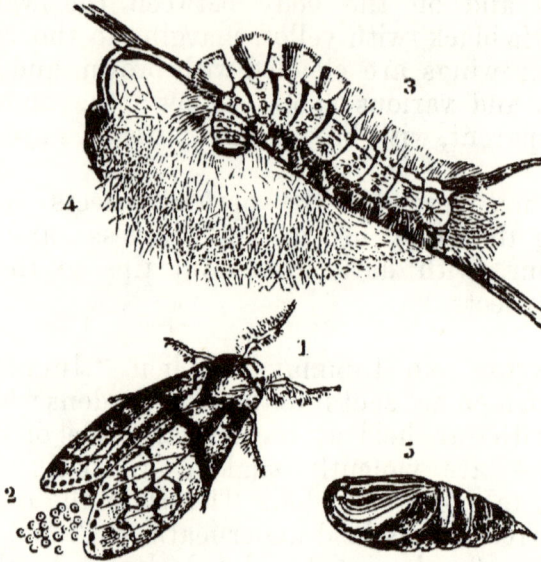

1, Female moth; 2, eggs; 3, caterpillar; 4, web; 5, chrysalis.

The moths are about two inches in the spread of the
fore wings; those of the male are grey, with a broad
smoke-coloured transverse bar, and brownish markings;
those of the female are paler grey, without the central
bar. The hind wings of both male and female are
nearly white, with a dusky streak near the hinder
margin. The head, body between the wings, and
abdomen are ashy white; horns whitish.—(E. N.,
J. F. S.)

PREVENTION AND REMEDIES.—If the caterpillars are in
any great quantity, a hard syringing is to be recommended; from the moisture lodging in its long hair,
heavy rain or any other application of water is a serious
annoyance to this caterpillar, and a help towards
diminishing its numbers.

From the size of the cocoons they are readily noticed, and all that are spun in leaves or on twigs may be easily cleared.

Foreign observations mention the caterpillars going down the stems in autumn to turn to chrysalids in moss, or at the surface of the ground; and it is recommended in that case to have them searched for and destroyed near infested trees. In the case of Hop-grounds this would be done in the regular course of cultivation.

Otter Moth. *Hepialus humuli*, Stephens.

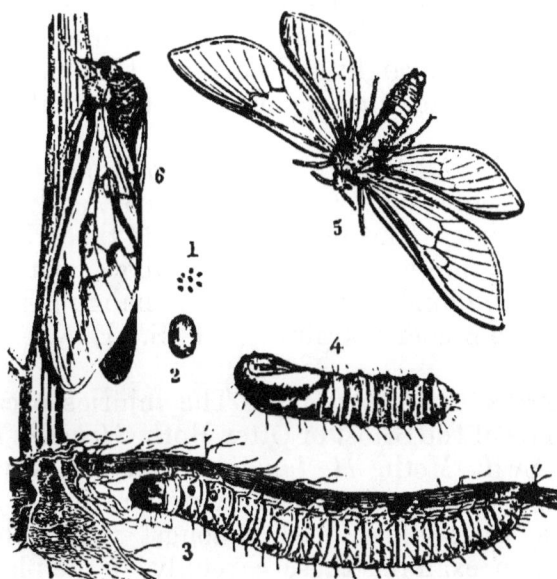

1 and 2, Eggs, nat. size and magnified; 3, caterpillar; 4, chrysalis; 5, male; 6, female.

The caterpillars of this moth (which is known also as the Ghost Moth, or Ghost Swift) injure the Hop by infesting the roots, "which they penetrate with their strong jaws, consuming the inside as well as the bark."

The moths fly in the evening, after lying concealed during the day amongst leaves or grass, and it is stated that during this flight the females drop their eggs one by one.—(E. L. T.)

The caterpillars are of a cream-colour, with brown heads, and a scaly patch on the next segment. They bury themselves in the ground, and feed below the surface until they are nearly an inch and a half or two inches long. When full-fed they spin a web amongst the roots on which they have been feeding, and in it they turn to stout, blunt, dark brown pupæ with two rows of spines.

This change takes place in May, and the moths are common in grassy places about the middle of June, especially in the south of England.

They are very peculiar in appearance, from the wings being somewhat straight and narrow; also there is not any great difference in size between the front and the hinder pair.

The male is about two inches in expanse of the wings, which are white above; the head, body, and abdomen are pale tawny. The female is about three inches in expanse: the fore wings are yellow above, with orange markings; the hinder wings smoky, changing to bright tawny on the hinder margin.—(J. F. S., E. N.)

PREVENTION AND REMEDIES.—The injuries occasioned by the larvæ of the Ghost or Otter Moth, *Hepialus humuli*, and the Swift Moth, *H. lupulinus* (large fleshy grubs which attack the roots), are fortunately rare.—(C. W.)

As the caterpillars feed amongst roots, the best remedy is to examine these carefully, if the plants are found to be flagging without obvious cause. From their large size they are easily seen, and can be taken out by hand and destroyed; and if there is reason to suppose them present, it is well to have a stock here and there examined throughout the ground when the plants are dressed in the spring.

Also as the moths frequent grassy places, and the caterpillars feed on the roots of the Burdock and of the Common Nettle, it would be very desirable to clear off these large weeds and also patches of neglected grass which give harbourage by day to the moths.

Hop-vine Snout Moth. { *Pyralis rostralis*, Linn. *Hypena rostralis*, Stephens.

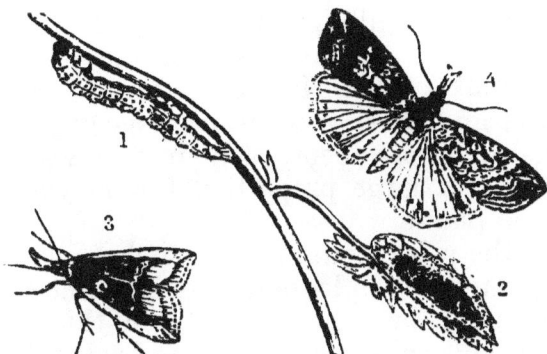

1, Caterpillar; 2, chrysalis; 3 and 4, moth.

The caterpillars of this moth injure the Hop by feeding on the leaves. They are of a pale green colour, with clearer spots, and a whitish line on the sides and back; slender in shape and gradually smaller towards the head, and of the size figured above; and are distinguishable by having only three pairs of sucker-feet in addition to the true feet on the rings next the head, and the pair of sucker-feet at the end of the tail.

When full-fed the caterpillar draws a leaf partly together, and commonly changes to the chrysalis in a light cocoon which it spins within the folds.

The moth, which appears in June and July, or earlier, is rather more than an inch in the spread of the wings, and is variable in colouring. In well-marked specimens the fore wings are greyish brown, darker from the base to about the middle, with a zigzag blackish streak across near the tip, and some raised tufts of black scales about the centre; the hinder part of the wings is palest. The hinder wings are brownish.

The moths of this genus may be generally known by the snout-like appearance of the front of the head, whence they take their name.—(' Illus. Brit. Entom.')

PREVENTION AND REMEDIES. — Strong syringing, by means of the engines, with some of the regular Hop-washes, or with the common fish-oil soft soap procurable at ten to twelve shillings per firkin of sixty pounds, appears to be one of the best remedies known. The soft soap makes the Hop-bines unpleasant to the caterpillars, and, if applied as soon as any number of the moths are seen about early in the season, would probably deter a large proportion from laying eggs on the syringed plants.

Striking the poles is also recommended, so as to make the caterpillars (which loose hold on a slight shake) fall to the ground; but the difficulty in this method of treatment is to keep them from going back again up the plants. Trampling on them, throwing soot, lime, &c., or, in bad cases of attack, shaking them into something placed below or on to tarred boards, are recommended.

Hand-picking the leaves with the moth-cocoons inside gets rid of much of the second brood, and these cocoons are also to be found about the stems of their various food-plants, on the surface of the ground, or in sheltered nooks; and clearing away plant-rubbish generally, and more especially the Nettle on which this caterpillar feeds, would be of service.

Red Spider. *Tetranychus telarius*, Linn.

The "Red Spider," which causes enormous damage in dry seasons to the Hop-crops, is neither an insect nor (properly speaking) a spider. From the first it may be distinguished at a glance, magnifier in hand (as shown by the figure on next page), and from true spiders it may be known by the body and abdomen being in one piece, and not merely joined by an almost thread-like connection. Strictly speaking, it belongs to the "Spinning Mites," and the figure gives a greatly-magnified view of the long stiff hairs, with globular formations at the tip, which are supposed to be of use in spreading their webs on the leaves.

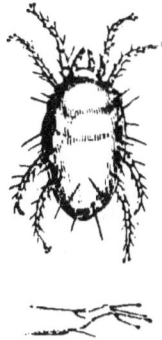

Red Spider.   Hairs on the foot (from Claparede), magnified.

These mites being scarcely discernible, excepting when collected together, and possible difference in species or in variety not affecting the matter of their prevention, I refer the reader, for their life-history and specific distinctions considered to exist by various writers, to the paper on 'Red Spider of the Lime Tree,' (? *T. telarius T. tiliarum*, Mull.), with figures of the mite, eggs, and webs, from life; and to 'Red Spider on the Plum.'

PREVENTION AND REMEDIES.—"This little mite, hardly to be distinguished without the aid of a glass, works much mischief in very hot dry seasons. Its effect upon the leaves of the Hop-plant was until recently attributed to heat and drought, and was called 'Fire-blast.' In the unusually hot and dry summer of 1868 the 'Red Spider' did immense damage in the English and German Hop-plantations. The leaves of the plants turned brown, became shrivelled, and fell off; and thousands of acres produced no hops.

"Mr. And. Murray, in his 'Handbook of Economic Entomology' (Aptera, pp. 88, 89), thus describes the

I

work of this Mite :—' On leaves (especially the under side of them) it finds a fit hold, and spins its web, affixing the threads to the prominences and hairs of the leaf; and under this shelter a colony, consisting of many of both sexes in maturity, and young in all their ages, feed and multiply with rapidity. The plant soon shows the influence of their presence in its sickly yellow hue; the sap is sucked by myriad insect-mouths from the vessels of the leaf, and its pores are choked by excremental fluids.'

"Washing the plants with soft soap and water, or even with pure water, is a remedy for these mites : in 1868 some planters tried a solution of sulphur thrown over the plants by the ordinary washing-engines, which killed the "Red Spider," but was done too late to save the crops; no doubt when hot and dry summers return again, washing with solution of sulphur will be largely adopted."—(C. W.)

The following note describes the progress of the attack, and the effect of rain in checking it; and also gives some idea of the amount of loss consequent on attack :—

"Red Spider in Hops. This only appears in dry seasons. It shows itself first by a brownish yellow appearance in the middle of the leaf, generally at the top of the poles, in large spots, spreading rapidly all over the Hop-garden, unless rain comes, which, if of any continuance, stops its progress. On looking under the discoloured leaves a small web will be found over the whole surface. After a time these leaves drop off, and the Hops all become red or brown.

"In 1868 I had fifteen hundredweight per acre reduced on several acres to five hundredweight, and then the five hundredweight were only worth about half-price. We do not use anything to stop it, because of the Hop being out."

"Red Spider" does not spread in late Hops which have been kept back by vermin, as the leaves are young and full of sap. It is dryness in everything that conduces to its spread, and I should think that this year one thousand acres of Hops have been left unpicked through it.—(J. W.)

The Red Spider appears to shelter itself in any convenient nooks—cracks in the Hop-poles, for instance, or, in winter, beneath clods of earth, stones, &c.; and all observations show the prevalence of heat and drought to be favourable to its increase; but whether these points can be met remuneratively by removal of sheltering-places, dressings of the ground, or applications of moisture beyond the customary washes, is still uncertain.

Hop Wireworm. } *Elater lineatus*, Linn.
Larva of } *Agriotes lineatus*, Esch.
Striped Click-beetle. }

Striped Click-beetle and larva (Wireworm), nat. size and magnified.

This species of Wireworm is the grub of the striped Click-beetle; it is considered to be probably only a variety of *A. obscurus*, but differs from it in having the wing-cases of the beetle marked lengthwise with greyish or dusky lines, whence the name of *lineatus*. It is stated by Bouché that the Wireworm of this kind (the *A. lineatus*) is sometimes to be found in great multitudes in dung, and in vegetable earth.

For life-history of "Wireworms," see this head amongst Corn-insects.

PREVENTION AND REMEDIES.—Wireworms are frequently very injurious to fresh-planted Hop-sets, especially upon recently broken-up pasture-land, eating off the tiny shoots directly they appear, and sucking the juices from the hearts of the sets. It sometimes happens that a large percentage of the sets die in consequence, and have to be replaced, involving great expense to the planter, as well as the loss of a year, The only efficacious way of getting rid of the Wireworm in a Hop-ground is to put traps of small slices of Mangolds, Potatoes, Carrots, or Swede Turnips, or small pieces of Rape-cake, close round the Hop-hills. These should be looked at twice a week, and the Wireworms which have eaten their way into them should be taken out and destroyed. As many as one hundred and fifty Wireworms have thus been trapped close to one Hop-hill.—(C. W.)

As these Wireworms, if once in possession, will live on in the same ground for several years (it is said five years), eating the whole time, excepting when they may go down deep in cold weather, it is a very important matter to save the expense of trapping or attack, by taking measures that the ground should be as far clear of them as is possible before planting, and also that they should not be brought in with vegetable-soil.

When pasture is to be broken up for Hops, it is of service to brush it early in autumn with chain or brush-harrows, and dress it with lime-compost; this is a great preventive of the beetles laying their eggs. Folding sheep so that the grass is eaten very close answers the same purpose.

In preparing the ground it is much better, as regards getting rid of the Wireworm, to trench with the spade two spits deep than to plough with the subsoil plough following. The weeds and grass-roots, and the like, in which the Wireworm feeds are not as thoroughly got rid of, even by deep ploughing, as by being thoroughly put down below by the spade, and the Wireworm has consequently plenty of food to keep it thriving until the new crop is put in.

Paring and burning is serviceable as a means of getting rid of the Wireworm and its food together, but has its drawbacks agriculturally; and it should be borne in mind that the Wireworm will go down as much as twelve inches in cold weather, therefore it is well to pare and burn before cold has set in ; also the parings should be collected and burnt at once, or the Wireworms will very speedily secure themselves again in the ground.

All possible care in removing the clods with grass-roots, and clearing the ground of rubbish which would keep the Wireworm in food-plants till the Hops come, would answer. (See " Wireworms" in Corn.)

The fresh vegetable-soil from old hedgerows or similar places brought in to restore old Hop-grounds is particularly likely to bring in Wireworm. It would save after expense to have the fresh field-soil heaped with lime or gas-lime, and the surface of the heap turned from time to time, to destroy grass and weeds which would, as in the field, feed the grubs. Gas-lime would thoroughly destroy all it touched—insect or plant—at first, if fresh from the works, but the atmospheric action or mixing with the earth would rapidly change its chemical nature to the sulphate of lime or gypsum serviceable to the crops.

It is noted that in planting, "one good well-rooted set in good soil will make as good a stock as two or more; but it is safer to put two, for fear of Wireworm" (C. W.); and in the early spring season, when the Hop-plants are dressed and the hills covered with a little fine earth, it would probably answer well to add some insect-deterrent.

For Wireworm in Turnips it has been found to succeed well to mix sand with just enough paraffin to moisten it slightly—not enough to clog, but still to run in the hand—and to sprinkle this *very* lightly over the roots by hand. Ashes or dry earth would answer almost as well as sand, and as the Wireworms usually feed near the surface, the smell or the paraffin in dilute state driven down by the rain would soon tell on them.

Rape-cake at the rate of five hundredweight per acre, crushed into lumps of about the size of half-inch ground bones, ploughed or harrowed well in, is stated to answer well; on the other hand, it is stated on good practical authority that Rape-cake is of no use as a preventive, but rather encourages the Wireworm.

LETTUCE.

Lettuce-root Aphis. *Aphis (Pemphigus) lactucæ*, Westwood.

These Aphides are found beneath the surface of the ground on the roots of Lettuce. If the plants droop suddenly without any evident cause, in warm weather, it is likely that on examination the roots may be found covered with small greenish plant-lice, which are destroying the plant by sucking away the juices.

This species is very small, only about the twelfth of an inch long, and scarcely a quarter of an inch in the spread of the wings, and is much like the common Aphides, the Green Fly of the Plum, for instance, but is distinguishable, with the help of a glass, as one of the *Pemphiginæ* by the third vein of the fore wings, counting from the body, *not* having any fork. The perfect insect is of a pale greenish yellow, with the head and body between the wings of a browner green; the pupa is also of a dusky or dirty pale green colour, with the tip of the body wrapped in cottony down.

This insect is without honey-tubes. — (J. O. W., in 'Gard. Chron. and Ag. Gazette.')

PREVENTION AND REMEDIES.—It is very hard to apply either, as, the Aphides being under ground, the first sign of their presence is the plant drooping from the extent of the injury, and then it is very likely too late to save it. Anything to be done to the attacked plants must then be done at once, and those near should be carefully examined by drawing the earth gently away for a few inches down the stems, to see whether there is any Aphis-attack beginning.

Drenching the ground round the Lettuces with strong soap-suds, soap-suds and tobacco-water, lime-water, tobacco-water (of a proper strength), or dilute Soluble Phenyle, may be good as direct remedies if applied at once; at least there does not appear to be any more hopeful method of treatment noted. The best treatment, however, would probably be to give good waterings of liquid manure, or dilute guano; this would be in every way beneficial; it would not suit the Aphides, and it would throw the plants into luxuriant growth and carry those that were only a little injured over the attack.

As matter of prevention, a dressing of sand or dry earth saturated with spirit of tar, and placed by hand round the neck of the plants, is recommended; also a good dusting of lime, or of soot over the ground, might be of service in making it very unattractive to the plant-lice to go down into.

The attacked plants should all be removed as soon as possible, in such a way as would destroy the Aphides, and neither leave them in the ground nor drop them all about to crawl or fly to the neighbouring plants. If the Lettuces are drawn, it would be desirable to cut off the root of each plant just below the collar (as it is pulled up), and drop it into a pail of any mixture that would kill the Aphides, or cripple them for flying; or the roots might be cut off into anything convenient and burnt; or, if the plants were waste, the whole might be thrown at once to the pigs. Any way in which the Aphides are surely destroyed will serve.

The infested spots should be purified by half a spadeful of quick-lime, or gas-lime, or something to kill the pests remaining in the ground where each Lettuce has been drawn, unless the bed is cleared completely, so as to allow regular measures of cultivation which would destroy the Aphides over the whole ground.

Lettuce Fly. *Anthomyia lactucæ*, Bouché.

The maggot of this Fly feeds on the seed of the Lettuce, and sometimes occurs in such quantities as to cause a complete failure of the seed-crop.

The Fly lays her eggs on some part of the flower (probably amongst the outer scales), and the little

maggots on being hatched gnaw their way into the base
of the flower and into the seed-grains. These grains
they completely clear out, so that only the outer husk
remains, and when the seed is eaten they leave the skin,
and go on to another seed-vessel or turn to pupæ in
the Lettuce-head, or in the ground if they have fallen
down.

The maggots are found in the latter part of the
summer, and are abundant in September. They are
much like those of the Onion and Cabbage Flies; legless,
tapering at the head, and cut off short and toothed at
the tail; yellowish white, and about the third of an
inch long.

They turn to pupæ in the autumn; these pupæ or
fly-cases are of a bright chestnut-colour and oval shape,
and (as seen through a magnifying glass) are rough.

The flies hatch in the following year, and appear from
April to June; they are very like the Onion Fly in shape
and size (see fig.), but somewhat different in colour.

The males are black and bristly, with face inclining
to chestnut-colour; four whitish stripes more or less
plain on the fore part of the body behind the head;
abdomen grey, each ring blackish at the base, and with
one black triangular spot; the legs are black; and the
two wings are stained with black.

The female is grey, with a bright chestnut stripe down
the face; blackish legs, wings, and nervures lighter than
those of the male.—('Gard. Chron. and Ag. Gazette.')

PREVENTION AND REMEDIES.—A great deal may be done
to prevent the appearance of this fly by examining the
Lettuce-seed carefully before sowing.

If the pupæ or little brown fly-cases are sown with the
seed, the fly will come out of them in due time and make
its way up through the earth, lay eggs, and so give rise
to an attack of the maggot. It is therefore recommended
that infested seed should be cleared in whatever way may
be preferred. Probably passing the whole through a
sieve with a mesh of such a size as would not let the

pupæ fall through would be the easiest method, taking care of course to burn or scald these pupæ directly.

In this case, however, the seedsmen are the proper parties to look to for remedy. More careful examination of the seed on receiving it from the producers would be necessary, and, if the seed was infested, measures might be adopted to clear it which, on this large scale, would probably be more effective than those the grower could, or would, in most cases adopt for the small quantity required for his crop.

It may also be worth consideration whether returning seed unfit for use to the dealer is not a desirable plan ; our leading seedsmen would wish to be able to correct their own suppliers, and, in the case of carelessness or wilful mis-serving, attention ought to be drawn to the matter.

Infested crops that are cleared off as worthless should of course be burnt at once ; not thrown in a heap with the pupæ still in them, to hatch next spring.

ONION.

Onion Fly. *Anthomyia ceparum.* Bouché.

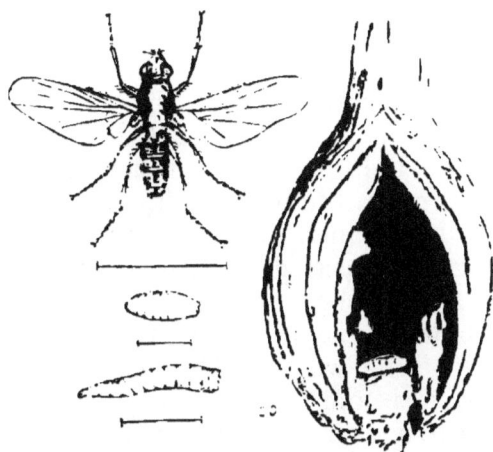

Onion Fly, pupa, and larva. all magnified.   Onion-bulb, showing pupa
remaining in stored Onion.

The injury in this case is caused by the maggots of
the Onion Fly feeding inside the Onion-bulbs, which,
partly from the quantity gnawed away and partly from
the decay caused by the workings, are often completely
destroyed.

These maggots may be found as early as May, whilst
the Onions are still small, and the attack is stated to
begin by the fly laying her eggs on the leaves of the
Onion close to the surface of the earth, from which point
the maggots make their way between the leaves into the
lowest part of the Onion-bulb, where they may be found in
numbers varying from two or three upwards. Here these
yellowish white legless maggots, of the shape shown
magnified in above figure, feed for a fortnight, then they
(usually) leave the bulb and enter the earth, and there
turn into chestnut-coloured pupæ or "fly-cases," formed
of the hardened skin of the maggot, of the oval shape
also figured above.

From these the fly comes out in from ten to twenty
days, in summer, and almost immediately lays her eggs,
and thus starts a new attack on the Onions which may
have escaped before; and so the destruction goes on as
long as any of the Onions remain and the warm weather
continues.

In the case of the later broods, however, I believe,
from my own observations, that the eggs of the fly,
instead of being laid on the leaves (which by that time
are far above the bulb), are laid either on the ground
close to the bulb or on the bulb itself. At that stage of
growth it is often much exposed, and in the bulbs
examined the grub appeared to have entered either quite
at the base or a little above it.—(Ed.)

The fly is figured magnified, with line showing nat.
size in spread of the wings. The male is ashy grey,
with black bristles and hairs; the face white, with
black horns; three dark lines along the body between
the wings, and a row of long blackish spots along the
abdomen; the female more ochreous, with yellowish white
face. The flies may be found throughout the summer,
but such of the maggots as turn to pupæ in the autumn
remain in that state till the following spring, and then
come out as flies in April or May.—('Gard. Chron. and
Ag. Gazette.')

PREVENTION AND REMEDIES. — The attacked Onions
decay, and may be known by their leaves fading and
turning yellow. These Onions should be cleared out of
the bed at once. This is a most important point, and,
if thoroughly done, will be likely to prevent all loss from
further attack, as thus all the maggots that would
develop into the next brood of flies are got rid of. But
these Onions should *not* be hand-drawn; they should be
taken up carefully by means of a spud, an old knife, or
some similar instrument, so as to get the whole of the
Onion and the maggots in it up together. The Onion
that is badly attacked becomes a rotten mass towards
the lower part, with the maggots inside just covered by

the skin of the bulb. If the leaves are pulled-at in the usual way they come up easily, but the decayed base of the bulb and maggots remain behind, and, though the bed looks better, no good at all is done. If carefully raised, each rotten Onion will lessen the next hatch of flies by the number of maggots contained in it. These should be put as they are drawn into a pail or vessel of any kind, out of which they cannot creep, and carefully destroyed.

The pupa or fly-case of the autumn brood remains in the ground, or sometimes in decayed Onions, during winter, and the fly comes out from it in time to attack the young Spring Onions; therefore Onions should not be grown two years running on the same ground. It has been observed that the fly, when buried a few inches deep, has much difficulty in coming up through the ground if it is at all firm. For this reason, when Onion-beds have been much infested, it is a good plan to deeply trench the ground, turning the top spit into the bottom of the trench. By this means, the fly-cases are buried so deep that the flies cannot come up from them, and also the cases are not brought up to the surface again by the common routine of digging or cultivation; it would probably answer every purpose, as far as keeping the fly from coming up is concerned, if the surface was simply turned down one spit deep, but it should always be borne in mind that room must be allowed on the top of the turned-down vermin for regular processes of cultivation to go on without reaching them, till after hatching-time. If ground with buried pupæ that naturally hatch in April is turned up again in March, no good will have been done by burying them.

It appears that the most successful method of culti-vation is, to trench the ground for the Onions in autumn, or at least early in winter, working plenty of manure into the soil, or placing a good layer at the bottom of the trench. Farm-yard manure, well-decayed, is suitable for this. In one locality, where the soil is heavy and damp, it is stated that horse-manure not too much fermented answers best.—(J. S.)

The soil is ridged or laid up rough for the winter, to expose it as much as possible to the air and frost.

The ground is prepared for sowing about the end of February, or as early in March as weather may permit, by levelling it with forks. Before or after this levelling (as may suit the views of the grower) some compost of a stimulating kind is added, and slight differences occur in details of successful treatment. In one locality, on the first favourable day in March, the ground is well-trodden, the seed is sown in drills, and before the drills are closed a compost is sown broadcast and pretty thickly over the ground, formed as follows:—Four parts refuse soil from the potting bench, two parts dry soot, and two parts wood-ashes; these are thoroughly mixed together, and put through a fine sieve before being used. —(F. G. F.) Soot and wood-ashes are also noted as applied before levelling for seed-sowing early in March. —(A. S.)

In another case about the 10th of March, if the weather is favourable, about six or seven cart-loads of hen-manure are wheeled on to the Onion-break previous to forking (care is taken to have the hen-manure well turned during the winter, and covered with soil to keep in the ammonia). After forking, the ground is well raked, the seed sown in rows a foot apart, and, after it is covered, the soil well trodden with the feet and raked over. The trampling of the feet makes it quite hard, and is considered to do good by preventing progress of the maggot.—(G. M'K.)

The following notes refer more especially to *light land*:—

The same kind of general treatment mentioned above, with some slight difference in detail, and successful on light land, consists in manuring with well-made manure, principally cow-manure. The ground is worked deeply, and trenched if necessary. After lying exposed to the frost, the surface is pulverized with a fork, but without turning up the manure, and then trodden well down before sowing, which is done as early in March as possible. The lines are a foot apart, and a heavy roller is passed a few times over the ground after the seed is covered in. The Onions are sowed thinly, so that the plants seldom require thinning.—(D. M.)

This method of treatment has been found very successful, and the point of thin sowing deserves particular attention.

With some of our crops the mere fact of thinning causes insect-attack, though whether this is from the bruised plants attracting the insects, or the check to their growth, or what the reason may be, is not yet clear.—(Ed.)

On a light sandy loam which had been exhausted by long use as garden ground, and the Onion-crop regularly failed, a dressing of clay fertilised by strong liquid manure was found to answer. In this case some rich clay was laid down in July in a place "where all the slops of the house" were thrown, and where it remained till the February following. The ground intended for Onions was dug in the autumn; in February it was lined into beds, and the manured clay was spread three inches thick on these beds, and left in this state till the second week in March; when it was broken smooth, and stirred, stirring some of the earth below with it. A good barrowful of pigeon's-dung (fresh from the pigeon-house) to every twenty yards was then spread on the surface, and the seed was sown, pressed hard with the back of a spade, and covered about a quarter of an inch deep with well-broken earth from the alleys. The Onions thus grown are reported as of excellent size and quality, whereas those treated in the usual way of manuring were nearly destroyed by maggot.—('Gard. Chron.')

With regard to special applications for prevention of the maggot, it is shown, by observations of 1879, that the use of paraffin-oil, either applied diluted with water or mixed with sand as a dressing on the beds, is very serviceable.

It is noted in one locality that, when the maggot appears, paraffin is mixed with water in the proportion

of a pint (English measure) to two gallons of water, and
with this the Onions that are planted in rows are
watered through the spout of the can without the rose;
the paraffin should be used carefully in dry weather,
lest it should burn the plants.

Another method of application is, to mix a good
glassful of paraffin-oil with about six gallons of water,
and throw this mixture carefully as a spray over the
Onion-beds; this cured the maggot-attack after two or
three applications.—(J. W.)

In another locality the only means found serviceable
for prevention of maggot-attack was the use of sand
saturated with paraffin-oil, and sprinkled amongst the
Onions, this sand being afterwards watered by means of
a can with a rose.   In this case experiment was made
as to the direct effect of paraffin on Onion-maggots, and
twenty-four hours after the application of three drops of
paraffin to the soil in a flower-pot containing some young
Onions and Onion-maggots, these maggots were found,
on examination through a magnifying glass, to be (with
the exception of two) all dead.—(G. M'K.)

Soap-suds are very useful; it has been found that
there is no trouble with the maggot where watering with
soap-suds is freely given on its first appearance; the
suds usually destroy the maggot in two or three applications, and also nourish the Onions.—(P. L.)

It is a good plan to pour the suds over the plant
through the rose of a watering-can, so as to make them
disagreeable to the fly, as well as effective to the grub in
the ground.

The ammoniacal matter contained in house-slops make
them a valuable manure for Onions.   They may be
applied with advantage to the growing plants, and, in
cottage gardens, good crops are taken off ground where
these slops have been thrown during the winter, and the
bed dug and sowed in the spring.

Liquid manure from farm-yard tanks, diluted with
water till reduced to a safe strength, has been found
useful; also the use of liquid drainings from pig-sties

have been observed to give good crops when all others in the neighbourhood have failed by maggot-attack.— (J. K.)

A heavy watering, to render the fertilizing matter in the soil available, is often of much service in running the growth on healthily without check in dry seasons.

Soot, charcoal dust, and pulverized gas-lime have all been found of good service in checking attack; but probably the use of ground clean from the Onion Fly to start with, and so prepared that a healthy vigorous growth is likely to take place, a watchful eye to remove infested plants as soon as they show, and the immediate application of a strong drenching of the fluid most obnoxious to the grub that can be found, are better than dressings not rapidly soluble, for the fluid goes down at once and lodges in the maggot-holes.

To the above I may add that the worst cases of maggot-attack I have seen followed sowing on ground which was not prepared beforehand, and had not received a special dressing of manure.

Guano is recommended by some growers, but I have known such severe attack of Onion Fly accompany its use as a dressing, that I have not mentioned it as desirable.—(Ed.)

K

PARSNIP.

Carrot and Parsnip Fly. "Rust." *Psila rosæ.*
Parsnip and Celery Leaf-miner. *Tephritis onopordinis.*
Carrot-blossom Moth. *Depressaria daucella.*
Carrot and Parsnip-seed Moth. *Depressaria depressella.*

I am not aware of the Parsnip being seriously injured in this country by any insect peculiar to itself. Its chief enemies are those above mentioned, which are noticed under the heads of insects infesting Carrots and Celery.

PARSNIP AND CELERY LEAF-MINER. *Tephritis onopordinis*, Curtis.

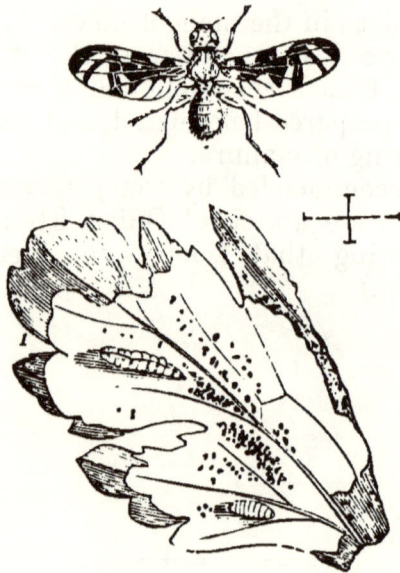

Parsnip and Celery Fly, magnified; line showing nat. size; larva and
pupa figured in blistered leaf.

PEAS.

Pea Moth. *Grapholitha pisana,* Curtis.

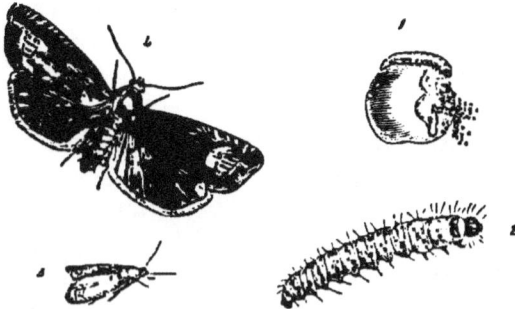

1 and 2, Caterpillars on Pea, and magnified; 3 and 4, moth, nat. size and magnified.

The caterpillars of this Moth cause the "worm-eaten" or "maggoty" Peas often found in old pods when the crop is maturing; the insides are eaten away, and filled with and partly surrounded by the excrement left by the caterpillar.

In the account of the method of attack of what I believe to be this Pea Moth, under the synonym of *Grapholitha nebritana,* Tr., it is stated by Herr E. L. Taschenberg that, at flowering-time, the Pea-plants often swarm with these little moths as soon as the sun has set; the females lay as many as three eggs on one young pod (or embryo seed-vessel) in the blossom; after about fourteen days the young caterpillars creep out of of the eggs, and pierce their way into the pod, where they feed on the seed.—(E. L. T.)

These caterpillars or maggots are fleshy, and slightly hairy; scarcely more than a quarter of an inch in length, and are generally yellowish in colour, with a black head, a brown band on the ring next to the head, and eight brown dots on most of the following rings. They sometimes, however, vary in colour; in some specimens the head and the next ring are brown, and in some they are intensely black. The legs on the three rings next to the head are black.

The caterpillars go down into the earth to change, where they spin a cocoon (that is, a kind of egg-shaped covering formed of silken threads drawn from the mouth) in which they remain till spring, when they turn to chrysalids, out of which the moths appear in June.

The moths are rather more than half an inch in the spread of the wings, satiny, and mouse-coloured. The upper wings have a row of very short white streaks directed backwards from the front edge, and have a silvery oval ring with five short black lines inside it placed near the hinder margin.—('Farm Insects,' and 'Praktische Insekten-kunde.')

PREVENTION AND REMEDIES.—In gardens where, as a rule, the greater part of the Peas are picked green, a large number of the Pea Maggots are destroyed in them, and a portion only are left to arrive at full development in the pods that may be left hanging. In such cases, thorough raking over of the ground, and deep-digging of the same before winter, may be expected to destroy most of those that have remained.—(E. L. T.)

In this way many of the maggots would be brought to the surface and cleared off by birds, and the operations of digging or trenching afterwards would bury a large proportion of the rest so deeply that, even if the caterpillar was not killed, the moth when developed would not be able to come up through the weight of earth.

Where maggot-attack is noticeable in the pods, the Pea-haulm should be cleared away directly the crop is gathered, so that all stray pods (which are very likely to be infested) may be cleared off the ground before the maggot can go down into it. This haulm should be carefully destroyed at once; the safest way is to burn it, but it may be thrown into the stock-yard, or buried beneath wet manure in the muck-heap; only it should not on any account whatever be thrown merely aside: the maggots will in that case go down into the ground beneath the rubbish-heap, whence the moths will come up again next year to infest the crops in the neighbourhood as before.

Alternation of crops, so that Peas should not be on the same ground, or close to the same ground on which they were grown the previous year, is a desirable means of prevention.

It is mentioned that the moths have been noticed swarming on the plants immediately after sunset; at this time there would often be dew, consequently any dressing thrown along the rows of plants would adhere, and any insect-deterrent chosen at the pleasure of the cultivator might do good service in keeping off attack. Quick-lime, soot, or any similar application, would be of use ; or hellebore-powder might be used in this case, as it would not come in contact with any part to be used as food, and even the pod containing the future crop would be at the time of attack still in embryo.

Pea and Bean Weevils. $\begin{cases} Situna\ lineata,\ \text{Linn.} \\ \quad,,\quad crinita,\ \text{Olivier.} \end{cases}$

These Beetles are often very injurious to the Peas, as well as to other leguminous crops, as Beans, Clover, &c. The attacked crops may be known by the leaves being scooped out at the edge, as figured. The beetles begin their work at the edges of the leaves, and gradually eat their way onwards, until, in bad attacks, nothing is left but the central rib, or merely the leaf-stalk.

The beetles appear about the end of March, and are numerous till May, when they may be observed pairing. The female is after this period full of white, somewhat transparent eggs, which, when in captivity, she deposits freely on the surface of earth, leaves, or even glass ; but I have never seen larvæ produced from eggs in these circumstances, and, as far as I am aware, the natural place of deposit of the eggs and the history of these weevils in their early stages is still unknown.

The Striped Pea-Weevil, *Sitona lineata*, is of an ochreous or light clay-colour, with three whitish or ochreous stripes along the back, and with ten punctured *stripes* alternately of a darker and lighter clay-colour along the wing-cases; the horns and legs are reddish.

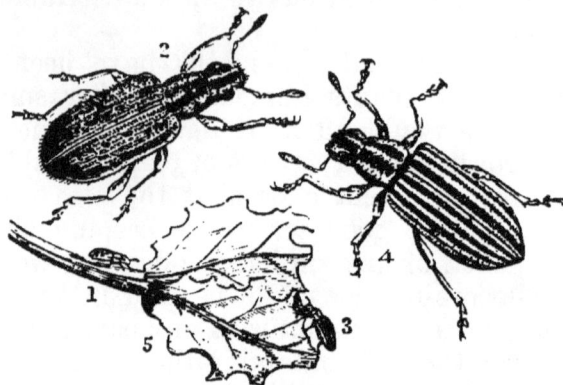

1 and 2, *S. crinita*, nat. size and magnified; 3 and 4, *S. lineata*, nat. size and magnified; 5, leaf notched by weevils.

The Spotted Pea-Weevil, *Sitona crinita*, differs from the above in being rather smaller, and more of a grey or rosy colour, with short hairs; and in the wing-cases, which have short bristly hairs down the furrows, being *spotted* with black.

The colours of the beetles are caused by the scales, with which they are thickly covered, and therefore only show well on fresh specimens; after a while they get rubbed off, and the black skin of the beetle appears in patches.

These weevils feed by day, and shelter themselves in the ground, under clods of earth or rubbish, at night. They may be found, according to season, and the crops there may be suitable for their food throughout the summer; but it needs some care in approach to see them on the plant, as they drop down at the vibration of a heavy step, and lie awhile as if dead.—('Farm Insects.')

PREVENTION AND REMEDIES. — Pea-crops suffer most from attacks of the weevil in their early stages of growth, as at this time the plants are tender and the leafage young, and therefore more liable to injury; also the number of beetles that would do but little harm to a fairly grown plant, soon destroy one with only a dozen or so of leaves. It should be kept in view that a stunted growth, whether caused by the nature or method of cultivation of the soil, or the character of the season, increases the evil by keeping the plants back for a longer time in this critical stage of growth.

To obviate these difficulties it is desirable to provide a good seed-bed, friable, sufficiently moist, and rich in available plant-food, which may be obtained to a certain extent by Peas following Cabbage or root-crops in the rotation.

In garden cultivation, besides the liberal supply of manure needed to run on a healthy growth, it has been found to answer well "to put a little broken turf and wood-ashes along the drill, sow the Peas on this, and cover them with a little more of the same."—(J. S.)

A good depth of coal-ashes, placed at sowing-time along the drills on a clayey loam, answered excellently; the Peas received all the air and moisture needed for germination, and came up luxuriantly, the roots soon taking good hold in the soil.—(Ed.)

This helps to obviate "caking" or "crust-hardening" over the sprouting Peas, a matter that no agricultural practice will quite counteract in the field where it is caused by the nature of the soil; something, however, may be done by careful stirring between the rows.

The attacks of the weevils are noted as being worst in dry weather, and (as they feed by day) good syringings with water or any addition thought fit, such as would make the plants distasteful to the beetles and encourage healthy growth, would be serviceable.

It is noted that a dressing of lime or soot given to the Peas (wetting them first to make it adhere) is an easily-applied and generally effective remedy.—(M. D.)

During the night the weevils shelter in the ground or under any convenient protection, so that any dressing which was obnoxious to the insects, but not bad for the plants, would be serviceable in diminishing the numbers that crossed it on their upward journey in the morning. Paraffin mixed with sand, mixtures of lime, gas-lime and soot, &c.,—as for Turnip Fly, for which see Index,— would probably be very useful; and well-prepared ground, free from clods of earth which may act as shelter to the beetles at night, or from any destructive application, is also desirable.

These beetles infest Bean as well as Pea plants, but the same treatment being applicable in either case, they are only noticed here under the head of " Pea and Bean Weevils," by which name (or shortly as " Pea Weevils ") they are described by John Curtis and by Prof. Westwood in their papers published in the ' Journal of the Royal Agricultural Society' and in the ' Gardener's Chronicle and Agricultural Gazette.'

POTATOES.

Colorado Beetle. *Doryphora decemlineata*, Say.

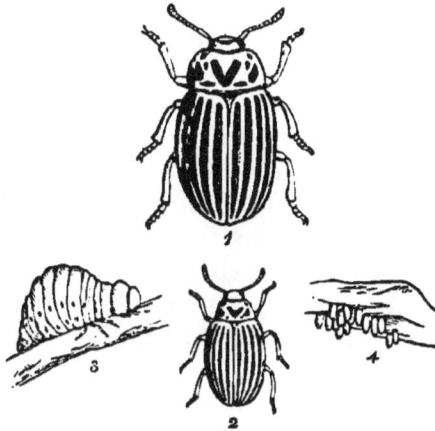

1 and 2, Colorado Beetle, magnified, and nat. size; 3, caterpillar; 4, eggs.

This Beetle is not mentioned here as a "British Injurious Insect," nor as one which (from the experience and knowledge of its life-history gained since 1877) can be deemed likely to be injurious to a serious extent in this country, but relatively to the precautions which it was thought desirable to adopt by our own Government, to prevent this fearful pest making good its footing amongst our Potatoes.

The eggs, figured above, are laid on the young shoots or beneath the leaves of the Potato; the grubs are orange or reddish, and change to pupæ in the ground; and the beetles are also distinguishable by their orange colour and by having (besides a large black spear-shaped mark on the back) *ten* black stripes on the wing-cases—five stripes upon each.

The natural home of this destructive beetle is in the Western States of America, and from Nebraska and Iowa it travelled eastward, until, in 1876, it reached the eastern shores of America; and in the autumn of 1877 specimens were found at Liverpool in a cattle-boat from Texas, which were identified by Mr. And. Murray (who was despatched to investigate the matter officially) as the "Colorado Beetle."

From what we now know of its life-history (as well as from the fact that, though nearly four years have passed since it was first certainly known to have appeared in this country, it has as yet made no settlement) we may reasonably believe that we are not likely to suffer from its ravages. Nevertheless it is a subject for prudent care, and, looking at the reprehensible custom that arose of sending living specimens in letters from America to this country by way of curiosities, we are probably greatly indebted to the Government restrictions *still in force* for saving us from constant difficulties and danger.

PREVENTION AND REMEDIES.—By an order in Council, published in the supplement to the 'London Gazette,' August 17, 1877, it is provided that—If the owner of, or any person having the charge of, any crop of Potatoes, or other vegetable, or substance, finds or knows to be found thereon the Colorado Beetle, in any stage of existence, he shall with all practicable speed give notice of the same to a constable of the Police establishment of the locality (the duties of the Police are unnecessary to be entered on) ; but it is further provided that it shall not be lawful for any person to sell, keep, or distribute living specimens of the Colorado Beetle in any stage, and any person failing to do anything he is required by this Order to do is for *each offence* liable to a penalty not exceeding *ten pounds*.

The surest remedy for attack is the use of "Paris Green," sometimes known as "Scheele's Green," a preparation of arsenic, and consequently a dangerous poison. This is procurable of all qualities at a price of sixpence per pound upwards, and "may be used in liquid suspension, in proportion of one tablespoonful of pure green to a bucketful of water"; and sprinkled over the plants by means of a watering-pot; a special apparatus arranged for carriage on a man's back; or a water-cart with tube and rose attached, or in many other ways, mixed with other substances, which it is unnecessary to enter on in detail.

Potato Frog Fly. *Eupteryx Solani,* Curtis.

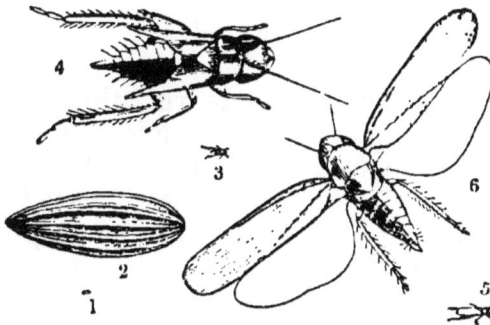

1 and 2, Eggs; 3 and 4, pupæ; 5 and 6, Frog Flies, nat. size and magnified.

This insect feeds in all its stages by inserting its sucker into the stem or leaf of the Potato, and drawing out the sap.

The eggs (figured above) are white and spindle-shaped, and are to be found upon the under side of the Potato-leaves.

The larva, which is much like the parent Frog Fly in shape, but without wings, is green when hatched, and is furnished with six legs, two horns, and a sucker; in the next stage (the pupa) it is green, nearly as large as the parents, but narrower; with black eyes, long black horns, and a stout sucker, by means of which it feeds until ready to change to the perfect insect. It then fixes itself firmly by its six legs to a stalk or leaf, the skin bursts along the back, and through the opening the perfect insect creeps out, leaving the deserted case standing as an empty insect-skin on the plant, or fallen beneath it. This "Frog Fly" is bright green, fading to a yellow tint, with short horns, brown eyes, and four iridescent wings less than a quarter of an inch in expanse; the upper pair glossy, somewhat rusty at the tips, and twice as long as the body; the lower pair exceedingly delicate. The length of the insect is about a twelfth of an inch.

The flies are very nearly allied to the common Cuckoo Spit Fly (*Tettigonia spumaria*), but they have not the power, like the Cuckoo Spit or "Froth Fly," of secreting a mass of frothy matter round them in their larval stages.—('Farm Insects.')

It does not appear that, as yet, this fly has been noted as causing any serious mischief to Potatoes needing prevention or remedy, but a short account of the insect is given here for reference in case it should be found that this species, as well as the *Eupteryx picta*, is one of the kinds of Frog Flies that at times do much damage to the Hops.

Death's-Head Moth. { *Sphinx atropos*, Linn.
{ *Acherontia atropos*, Curtis.

The caterpillar of this moth is sometimes found in great numbers feeding on Potato-leaves, but it seldom does any serious amount of damage.

When full grown these caterpillars are of great size, sometimes measuring four or five inches in length. They are thick and fleshy, with a pair of feet on each of the three segments behind the head, four pairs of sucker-feet, and another pair set close together beneath the last segment which act as claspers. Above this pair, on the back of the caterpillar, is a protuberance like a tail or horn, tubercled, and bent downwards, but turned up again at the tip. The head is horny, and furnished with strong jaws. The colour is generally yellow, or greenish yellow, speckled with black on the back, with seven slanting stripes of blue or lilac on each side; the upper end of these stripes form a kind of row of points, where they meet along the back, the lower end points forward, and is white or bordered by a white line. When about to change, the caterpillar turns to a lurid yellowish or reddish tint. It then goes down into the earth, throws off its skin, and turns to a large chestnut-coloured chrysalis. The caterpillars that change in July come out as moths in September and October; those that change in the autumn do not come out till the following spring.

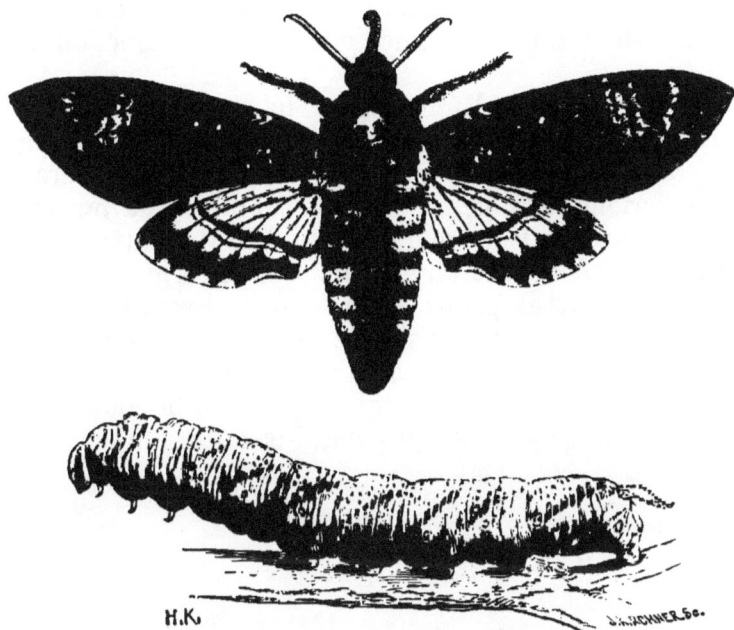

Death's-Head Moth and Caterpillar.

The moth is the largest of the British kinds. The spread of the wings is from four to five or even to six inches. The fore wings are of a rich brown, varied with yellowish or rusty tints, with black lines or cross-bands much waved and zigzagged; they have a pale or ochre-coloured spot in the middle, and are sprinkled with small white dots. The hind wings have the margin slightly scooped out, and are bright orange, with two brown or black bands, the outermost being broadest.

The head is black; the back has markings, in its thick black velvety down of pale ochre and orange colour, exactly like a painting of a skull or death's head, whence the name of the moth. The abdomen is yellow, with six black bands across; and a line or row of spots down the centre, and the tip of the tail are of a bluish grey colour.

Where the eggs are laid does not appear to be noted.

When the moth is alarmed it makes a sharp squeaking plaintive cry, not unlike the squeak of a mouse. It is said to enter Bee-hives and rob the Bees of their honey, whence the name sometimes given of Bee Tiger-Moth.

PREVENTION AND REMEDIES.—The caterpillar usually hides by day, and comes out in the evening or at night to feed. If it should occur in sufficient numbers to cause serious damage, it would be well for the owner of the Potatoes, or some one interested in the matter, to go through the field carefully at different times and ascertain when these great grubs are feeding. From their large size they are distinguishable in the twilight of the evening, or in a clear moonlight, so that, when it has been made out at what time they are to be found, they might be easily got rid of by hand-picking.

TURNIPS.

Turnip Aphis. { *Aphis rapæ*, Curtis.
                  ,, *floris-rapæ*, Curtis.
    ,,   Green Fly. { *Rhopalosiphum dianthi*, Schrank.

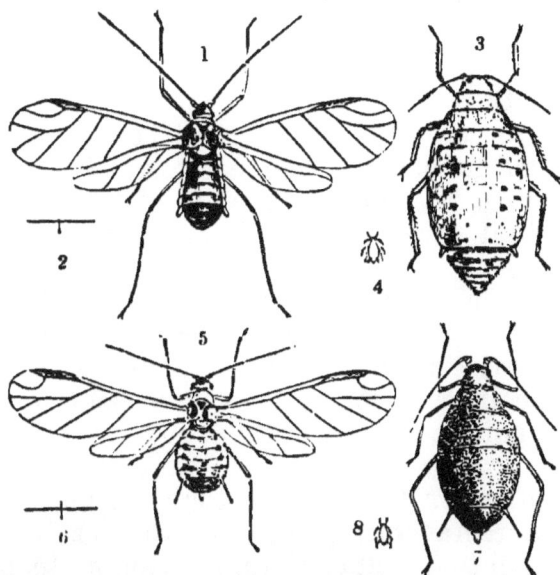

1—4, *Aphis floris-rapæ*; 5—8, *Aphis rapæ*, nat. size and magnified.

This Aphis is common in the summer on many kinds of plants, and is especially hurtful to Potatoes, Turnips, and Swedes. It is chiefly to be found on the under side of the leaves, but sometimes occurs in such numbers as to smother the plants; and the vast swarms of Green Fly which at times fill the air, as in the autumn of 1834 and in 1868, are believed to have been of this species.

The Aphis is very variable in appearance; the wingless viviparous female is usually of some shade of green, but often of an ochreous-red in autumn; the winged female is usually black, with reddish yellow abdomen, striped and spotted with black; ochreous legs; and wings yellow at the base, with yellow main vein; the insect is,

however, sometimes entirely black, ochreous, or green.
The variety *floris-rapæ* that is mentioned by Curtis as
found on flower-stalks of the Turnip, is described by him
as dull pale green, dusted with white.—('Mon. of Brit.
Aphides'; 'Farm Insects.')

PREVENTION AND REMEDIES. — In garden cultivation,
some good may be done by copious syringings with such
applications as ammoniacal water, tobacco-water, and
soft soap, the strength of the application being pro-
portioned to what the leafage will bear. One part of
gas-water or ammoniacal liquor to ten or twelve of water
will kill Green Fly, but the strength of the liquor varies
so much that experiment is necessary to ascertain its
power.

A proportion of twenty-eight pounds of soft soap and
half a pound of tobacco are used with one hundred
gallons of water at an expense of £2 2s. per acre as an
Aphis-wash in the Hop-gardens, this being applied by
means of garden-engines with double hose, worked by
three men. How far under the present arrangements
of field cultivation the mixture could be applied to the
crop at a remunerative cost, or whether it could be
applied at all to a field of Turnips, remains to be seen;
but if it could be brought to bear it would be likely to be
beneficial.

One great difficulty in counteracting Aphis-attack
arises from the skin of the Aphis being often of such a
nature, or covered with a mealy secretion of such a
nature, as repels water, and consequently many of the
applications simply run off them, without doing us any
good. Such poisonous applications as will kill them,
either by caustic action at once, or lodging where they
must be imbibed, are of service; and solutions of
common salt and of sulphate of soda have been stated to
be serviceable, but I have no details of proportion or
precise results, and, generally speaking, an adhesive
medium like soft soap, which may also be a vehicle
of some Aphis-poison, is the surest cure.

Fish-oil soft soap, expressly manufactured for application in field cultivation, is procurable from the City Soap Works, London, at the price of ten to twelve shillings per firkin of sixty pounds.

Aphides multiply most quickly in dry weather, and on plants which are sickly from drought, exhaustion by insect-attack, or other causes; so that all measures of cultivation tending to produce vigorous healthy growth are serviceable in counteracting attack; and where circumstances allow of the application of liquid manure, or of water to an extent to make the plant-food in the soil available, and push on growth that otherwise was being checked by drought, such treatment would be desirable.

Where new growth is not being made, and the juices are being constantly abstracted by the Green Fly, the plant necessarily fails, unless extra food is supplied to start it forward.

When Aphis-attack is found to be beginning in garden cultivation, it will be of service to break off infested leaves, or remove the infested plants, taking care to crush the leaves under foot, or destroy the Aphides at once in any convenient way.

Various kinds of Titmice, and especially the Blue Tit, are of service in destroying Aphides; and the common Ladybird Beetles and their larvæ feed on them voraciously.

There appears to be no further remedy for this and other insects that prey on the Turnip-crop when advanced to maturity than eating off the crop by folding sheep upon it: their treading and urine would exterminate everything, and thus be of some service as a means of prevention for next season.

Turnip Fly. $\Big\{$ *Phyllotreta nemorum*, Chevrolat.
Flea Beetle. $\Big\{$ *Haltica nemorum*, Curtis.

This is one of our most destructive kinds of farm-insects. With regard to its appearance little need be said; we all know the small black shiny beetles to be

L

found in myriads, in fine sunny weather, gnawing holes
in the Turnip-leaves as long as they are undisturbed,
but skipping off as briskly as the fleas from which they
take one of their names on being meddled with.

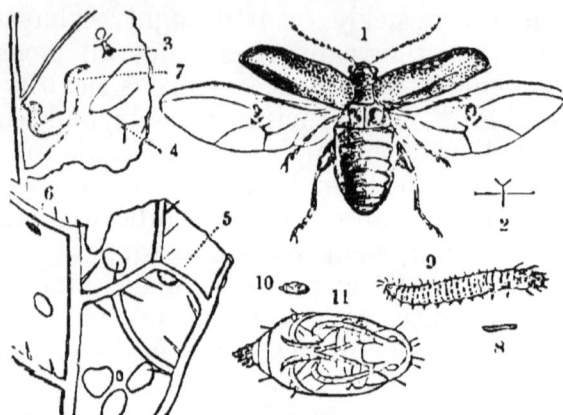

1—3, *H. nemorum;* 4 and 5, eggs; 6—9, maggot; 10 and 11, pupa;
all nat. size and magnified.

When viewed in the hand or under a magnifying glass
some difference may be found in their size and marking.
One of the largest and commonest kinds (*II. nemorum*)
has a broad yellow stripe down each wing-case; another
kind (*II. concinna*) is brassy, with a tooth on the second
and hinder pairs of legs; another kind is black, and
dark blue above; another is of a brighter blue above;
but these are (as far as we know) alike in their method
of living, in the harm they do, and the way in which
they do it. The same methods of prevention or cure
apply alike to all; and as it is almost impossible to
distinguish them unless they are caught, and also as,
however much they may vary in marking, still they are
all "Turnip Fly," which is the point of view in which we
are concerned with them here. I have only named two
species, and refer the reader who wishes for further
information regarding the different kinds to the 'Farm
Insects' of John Curtis.

During winter the Turnip Fly Beetles may be found sheltered under bark, fallen leaves, clods of earth, and the like places ; also amongst stubble, and especially in heaps of long strawy manure left on the fields ; and on particularly fine days they may be seen coming out to sun themselves.

On the return of spring warmth they begin work, and, till the crops are ready for them, are especially to be found on weeds of the same family as the Turnip and Cabbage, such as Charlock, Shepherd's Purse, and Jack-by-the-Hedge.

When the attack begins on the Turnip the female lays her eggs, which are few in number, for successive days on the under side of the rough leaf. The maggots, which hatch from these in ten days, are white or yellowish, fleshy, and cylindrical ; with three pairs of feet in front, and a sucker-foot at the end of the tail. The head is furnished with cutting jaws, and has large dark eyes. Directly they are hatched they gnaw through the lower skin into the pulp of the leaf, and make their way onwards, forming winding burrows inside it. Here they feed for about six days, then they come out and bury themselves (keeping near the Turnip) not quite two inches deep in the ground, when they turn into the chrysalis stage, from which the "Turnip Fly" or "Flea Beetle" comes up in about fourteen days.

It is in this state that the so-called "Fly" does most mischief. It gnaws the seed-leaves, and the young plant when it first springs, and thus often totally destroys it ; and also gnaws the rough leaves, forming large holes through the leaf.

There may be five or six broods in a season.—('Farm Insects.')

PREVENTION AND REMEDIES. — The points that need particular attention are, 1st, clearing off such weeds as the fly feeds on till the Turnips are ready for it ; 2nd, such a method of cultivation and manuring as shall give a fine, deep, clean, and moist seed-bed, rich in available plant food, so that a healthy and rapid growth may be promoted and all points of shelter or harbourage for the "Fly" be reduced to the lowest limit; 3rd, available means of applying moisture in dry seasons; 4th, applications and special treatment to destroy the fly when it is badly infesting a crop.

With regard to weeds:—The fly frequents wild plants of the Cabbage tribe, as Shepherd's Purse, Jack-by-the-Hedge, and is especially fond of Charlock. It has been observed as unusually numerous where this weed has been plentiful in the previous year, and also to spread (as from a centre) to the neighbouring crops from a Charlock-infested field. It is often supported in the spring by these or other weeds till the Turnips are large enough for it to attack, and therefore means should be taken to get rid of them beforehand from the autumn stubbles. In the case of Charlock a double turn of the harrow over the stubble is of use, small weeds may be cleared by broad-sharing; the seeds are thus covered sufficiently to induce immediate germination, and the sprouting weeds as well as roots in the soil will be cleaned by the regular processes of cultivation further on. Waste spots of land and hedge-sides should also be attended to; the first is often overrun with Shepherd's Purse; the second is often infested with the tall, large-leaved, onion-like smelling plant with white flowers, the shape of the Charlock-blossom, known as "Jack-by-the-Hedge."

A deep cultivation that will turn down weeds and destroy insects is very serviceable, and care should be taken that all manure from the yards or sheds should be completely buried. Any long strawy lumps left on the surface will shelter the fly, and from these it will come out to the destruction of the crop.

Plentiful rolling, harrowing, and other means of securing a fine seed-bed are also desirable, both that the ground may be in a state to push on a good growth, and also that the surface may be free of harbourage.

"Where clods prevail,
Turnips fail."

The three requisites for healthy germination are, warmth, moisture, and some amount of air; and it is only by securing these that a rapid and healthy development of the plant can be obtained. It has therefore been recommended, when the surface is prepared for drilling, to leave it undisturbed for three weeks; also, on the other hand, when partly-rotted farm-manure is ploughed-in in spring, to sow immediately. In each case the reason is the same, that is, to secure the moisture in the ground; in one instance by not opening the pulverized earth more than can be helped, and in the other by putting the seed above the half-rotted dung before the moisture and warmth accompanying decomposition has gone from it.

The Turnip requires moisture, and it is of great importance to run on a healthy growth by furnishing it with a sufficient supply, joined to a liberal application of ammoniacal and phosphate manures, as well as farm-yard manure.

For these reasons watering the ground in garden cultivation, and the use of the water (or liquid manure) drill in dry weather is very serviceable. This makes the food round the plant available at once, and where the seed is put in with artificial manure, as "Peruvian guano," superphosphate of lime, or whatever may be known to answer, gives it a good start.

The first difficulty in Turnip culture is the fly eating up the crop bodily whilst still in the seed-leaf, therefore anything that will run it quickly and healthily through this stage is a point gained; nothing will do this better than a well-prepared soil, with sufficient moisture to enable the Turnips to have the benefit of it.

Thick sowing up to eight or ten pounds per acre is advised by various growers, who state that thus, in case of hot dry weather, the plants will thrive better for the protection they give to each other (being thus moderately damp, with the roots shaded), and that some may be reckoned on to escape the fly. This, however, needs careful looking to, or the result will only be a worthless drawn growth.

The use of the horse-hoe is advised both before and after the seed comes up, and its frequent use afterwards is strongly insisted on.

Where water can be supplied in a season of drought to the young growing crop by a water-cart, or any other means, it will be very serviceable. In an instance noted where the plant was thus saved, the cost was fifteen shillings per acre. Here there was water at both ends of the field.

In garden cultivation watering would probably always pay; in field cultivation this would depend on local facilities.

With regard to Swedes, it has been found, from the preference of the fly for the White Turnip, that if the seed is mixed in the proportion of one quarter White to three-quarters Swede, or again, if one drill of White is put in at intervals amongst the Swedes, that the fly will be attracted to the White, and thus allow the Swedes to get well ahead. This plan was found to answer well by several years' experience in East Lothian and elsewhere.

With regard to direct applications to destroy the fly, it should be borne in mind that very much depends on the time of day and the state of the weather when these applications are made.

The Turnip "Fly" is active in bright dry weather; and when the thermometer stands at 75° in the shade, it has been observed on the wing in great numbers: when the weather, on the contrary, is cold and wet, it is sluggish; and in rain or heavy dew these beetles cannot leap, from the moisture clogging their legs, and thus preventing the powerful springs with which they customarily leap out of the way of attack.

This circumstance has much to do with the very different success, in different circumstances, of exactly the same remedy. A dressing that is put on early in the morning, whilst the dew is still heavy on the plant, has a very different effect to what it has either on a morning that is dewless or in the middle of the day, when the fly has every chance to protect itself under clods of earth, &c., before the dressing reaches it; and, though the reason is not given, the advice is constantly the same in observations on remedies—*apply whilst the dew is on.*

Rolling with a light roller as early as 2 or 3 o'clock on a dewy morning has been found to succeed well in bad attack during very hot and sunny weather.

Where the nature and condition of the soil permits, it has been found a good remedy, in attack of "Fly" as well as of Turnip Sawfly, to drive a flock of sheep over the attacked field very early in the morning, whilst the dew is still on the leaf. The sheep must be kept in motion, that they may not feed on the Turnips, and well up in a body, so as to tread the whole of the ground, which should be taken on successive days where there is a large extent to avoid injury to the sheep, as they must be kept moving. In one case noted the extent of ground was thirty-seven acres, and from four to five hundred sheep were driven over; and the cure of a bad attack was satisfactory.—(R. T.) The locality was a high-lying spot, with warm exposure in a chalk district, and probably it is only on a light or sandy soil that this remedy is applicable; on clay soils, or such as approach to a plastic character, it would be hazardous, if not simply destructive.

With regard to dressings when the fly is present:— Soot, lime, road-dust, and others of the usual applications have been found useful, and may all be serviceable if applied when the dew is on; but the remedy that appears the best proved is the one noted by Mr. Fisher Hobbs as having never failed during the eight years in which he made use of it. I give the recipe and passage at length from his statement made before the Council of the Royal Agricultural Society, quoted in the 'Gardeners' Chronicle and Agricultural Gazette' for May 28, 1859, p. 473:—

"One bushel of white gas-ashes" (gas-lime) "fresh from the gas-house, one bushel of fresh lime from the kiln,

six pounds of sulphur, and ten pounds of soot, well
mixed together and got to as fine a powder as possible,
so that it may adhere to the young plant. The above is
sufficient for two acres, when drilled at twenty-seven
inches. It should be applied very early in the morning
when the dew is on the leaf, a broadcast machine being
the most expeditious mode of distributing it ; or it may
be sprinkled with the hand carefully over the rows. If
the fly continues troublesome, the process should be
repeated ; by this means two hundred to two hundred
and twenty acres of Turnips, Swedes and Rape have
been grown on my farm annually for eight or nine years
without a rod of ground losing plants. The above is a
strong dressing to be used when the fly is very numerous,
and has never failed when applied at night. Numerous
experiments have been tried, and amongst them I
recommend the following in ordinary cases. . . . . .
Fourteen pounds of sulphur, one bushel of fresh lime,
and two bushels of road-scrapings per acre, mixed
together a few days before it is used, and applied at
night, either by means of a small drill or strewed along
the rows by hand. I have known sulphur mixed with
water applied in a liquid state by means of water-
carts during the night, and the horse-hoe immediately
following the water-cart. This has succeeded admirably."
—(F. H.)

The quantity that has been written on the matter of
prevention of fly is enormous, but all the notes show
that care should be directed especially to, 1st, cleaning
the ground ; 2nd, preparing, by good cultivation and
plenty of manure, to push on the growth healthily ;
3rd, sufficient moisture, and a supply of water (where
this can reasonably be given), in case of the weather
being too dry for the Turnip-growth to run on properly
without ; 4th, all dustings, dressings, &c., should take
place when the dew is on the leaf, and the fly exposed to
them, *not* in bright sunshine.

For further information, the reader is referred to the
papers from which this sketch has mainly been taken,

namely, the 'Farm Insects' of John Curtis: a paper "On the use of the Water Drill," by A. S. Ruston, 'Journal of the Royal Agricultural Society,' 1859; the paper above-mentioned by Mr. Fisher Hobbs; and observations recorded by agriculturists too numerous to quote individually, given in the 'Gardeners' Chronicle and Agricultural Gazette,' and also by observers recorded in the yearly 'Reports on Injurious Insects.'

Turnip Leaf-miners.

Black Leaf-miner. *Phytomyza nigricornis*, Macquart.
Yellow   ,,   *Drosophila ? flava*, Fallen.

P. nigricornis, nat. size and magnified; pupa; and leaf, with minings.

The Turnip-leaf is attacked by the maggots of two kinds of two-winged flies (*Diptera*), which, from their method of clearing away the substance of the leaves in galleries, are known as miners.

The Black-horned Leaf-miner, figured above, is a very small fly, only about the twelfth of an inch long, of a slate black or ash-colour, with an ochreous head, dark horns, and wings pale slate-colour; its maggot works beneath the skin of the under side of the Turnip-leaf so regularly that tracks are not seen on the upper side. The grub changed to pupa in the leaf, from which the fly came out in a few days, in the instance noted.

The other Leaf-miner is also a small two-winged fly, about the twelfth of an inch long—scientifically a *Drosophila;* the colour is generally ochreous, the face white, and the wings yellowish. The maggot is of a pale green colour, and is as constant in its habit of burrowing under the skin of the upper side of the Turnip-leaf as the other is in keeping to the under side.

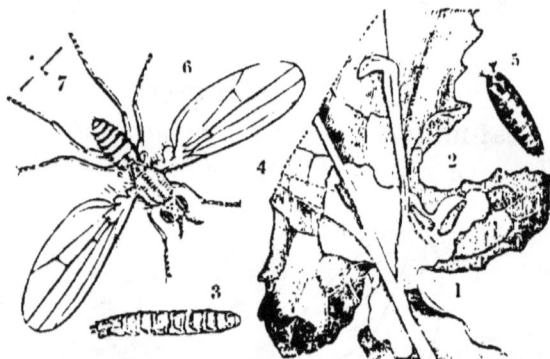

D. ? flava.—1, blistered leaf; 2 and 3, maggot; 4 and 5, pupa; 6 and 7, fly, nat. size and magnified.

It also differs (or occasionally differs) in leaving the leaf when full fed, and either remaining on the ground or going down into it to turn to pupa. These pupa or fly-cases are in both instances brown or chestnut-colour. The blisters are disfiguring to the leaves, but are not noted as ever occurring in such quantities as to cause any serious damage requiring prevention or remedy.—('Farm Insects.')

Turnip Moth.
Common Dart Moth. } *Noctua (Agrotis) segetum,* Ochsen.

The caterpillars of this Moth are destructive to Turnips, Cabbage, Mangold Wurzel, Carrots, Radish, Corn, Grasses, &c.; in fact they seem to feed on any plant that is not too hard for them to gnaw.

The egg is laid during the summer, as early as June or towards autumn, and the caterpillars hatch in about a fortnight.

These, when full grown, are about an inch or an inch and a half long, nearly as thick as a goose-quill, and smooth, with a few hairs; of a pale smoky colour, but sometimes pinkish, or purplish brown, and with two dark lines along the back and one along each side; these lines, however, are not always distinct. The head is a pale dingy brown, much narrower than the next ring, and is stretched out on a plane with the body.

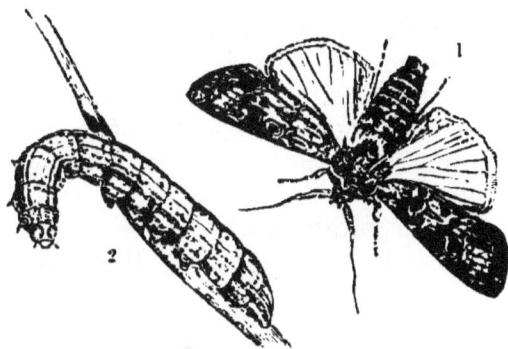

Moth and caterpillar.

When first hatched, the caterpillars appear to feed chiefly above ground, choosing the part of the plant just at the surface of the earth (between the root and stem), and, thus gnawing off the tops, they destroy the crop to a serious extent; as they grow stronger they go further down, and generally remain wholly underground or only come up at night to feed. Whilst the plants are young the caterpillars feed on those near them, going on to others as food gets short, gnawing off the tops as above mentioned, or feeding on the leaves, which, after having cut through the leaf-stalks, they drag partly down into their burrows, to be eaten during the day. When the Turnips are formed, the caterpillars establish themselves inside the bulbs, and as many as twelve may be found in one Turnip; these gnaw large holes and cavities, sometimes going completely through from one side to the other, and continue to feed there till the bulb is consumed, or till they have to leave it on account of frost or some other cause.

According to circumstances of climate, &c., they feed during the winter, or pass it in cells formed in the earth, coming out to feed again in the early spring. In May or June they turn to smooth brown chrysalids in the ground, from which the moth appears in about a month.

This moth has the fore wings of a pale grey ground colour in the male, dark umber-brown in the female, with various markings, as figured; the hind wings are pearly white, clouded towards the hinder edge in the female, and with dark rays. The colour of the body (including the abdomen) varies, like the ground colour of the fore wings, with the sex; it is lighter in the males than in the females.

It is not fully ascertained whether there are one or two broods of this moth in the year.—('Brit. Moths,' 'Farm Insects,' &c.)

PREVENTION AND REMEDIES.—These are not easy to find. In garden cultivation something may be done, where a plant is found to have had its head gnawed off, by turning up the soil close to it, and destroying the caterpillar, which will very likely be found an inch or two below the surface. If not removed, it will certainly go on and destroy another or more plants the following night, and continue so doing; so that the work of removal will probably pay.

It is stated to be a regular part of cultivation in some of our Cabbage-growing districts, in attacks of this kind, to put on a number of workers to dig down round the suspected plants, and thus turn out the grubs and destroy them.

Watering the ground with any special application does not seem of use, as the grub is perfectly well able to go down deeper, and so get out of the way of annoyance.

Tobacco-water is, however, said to kill all the caterpillars that it touches, and if this is the case, it might be useful as a watering applied overhead to the plants (of a strength tested relatively to safety of the foliage) at night, as soon as the caterpillars are out at feed on the leaves.

In the case of Cabbages where the roots were attacked, the treatment of applying a handful of soot round the stem of each of the plants and earthing them up immediately has saved the crop, and been followed by a good growth; this would do good by encouraging the growth when only a part of the roots had been hurt, and it would be very likely to keep the caterpillar from coming up through it to attack the plant.

We thus have some hold over night-feeding grubs, for (when the plant is once clear of them) if we can find anything to put round the stem that they will not come *up through*, or walk *over*, and that will not hurt the plant, we can keep it clear from their attack.

Gas-lime would do a great deal of good in this way. It is an excellent means of keeping off attack, and when put carefully by hand in a narrow ring round the stem of an infested plant has been found to answer well. The amount that is safe would vary with the age of the plant and time of exposure of the gas-lime to the air, but in the case of half-grown kale and gas-lime about two months from the works, a ring about as thick as a finger round (but not against the stem) is quite safe.

The mixture of gas-lime, &c., found serviceable by Mr. Fisher Hobbs as a remedy for Turnip Fly (see p. 151) would also be worth trying as a remedy for the " surface caterpillars."

Something may be done by hand-picking the caterpillars that are hidden under clods of earth, stones, &c., during the day, and the operation has been recommended for clearing the plants at night; but this is so very troublesome it seems a rather doubtful expedient.

With regard to field cultivation, as the caterpillars feed on almost all plants that are soft enough to gnaw,

and are voracious eaters, it is of great service to clear
the ground of all plants some time before the new crop
is sown. In districts where catch-cropping prevails, the
land ought to be ploughed and cultivated so as to be
cleared of all food for these grubs for (say) a fortnight
before the Turnips are sown; the caterpillars will thus
(as far as observations go) have either been literally
starved out or—as they are very active when in search of
food—will have strayed away to find it elsewhere.

The chief means of prevention, however, with this
caterpillar appears to be the encouragement of the birds
that especially feed on it. The Crow, Raven, Jackdaw,
and Magpie are all said to be useful, but the great
helpers are the Rook and the Partridge. The Rook
works down into the earth with his bill, the Partridge
turns out the grub by scratching, and where (as well
pointed out by the late Edward Newman) we increase
the quantity of any special crop so as to attract any
special insect, and at the same time allow the birds
which feed on those insects to be destroyed, we can
hardly fail to suffer severely. — ('Brit. Moths,' 'Farm
Insects,' Ed., &c.)

Diamond-back Turnip Moth. { *Cerostoma xylostella*, Curtis.
Plutella xylostella, Doubleday.
,, (syn.) *cruciferarum*.

The caterpillar of this Moth seldom does much damage,
but in 1851 it appeared in enormous quantities in
England and Ireland, in some cases almost clearing the
attacked crop. It feeds on the leaves of Turnips and
Swedes, and occasionally on Cabbage. This caterpillar
is about half an inch long and spindle-shaped (that is,
gradually tapering towards the head and tail), of a pale
green colour, with the head rather yellower, or grey; it
has a pair of short-jointed feet on each of the three rings
next to the head, and the foremost of these rings has a

number of small black spots; the two next have each two yellowish spots on the back.

As many as two hundred and forty of these caterpillars have been counted on a single plant of moderate size. They gnaw the leaves away down to the veins, and, where the attack is bad, clear off these afterwards so as utterly to destroy the crop.

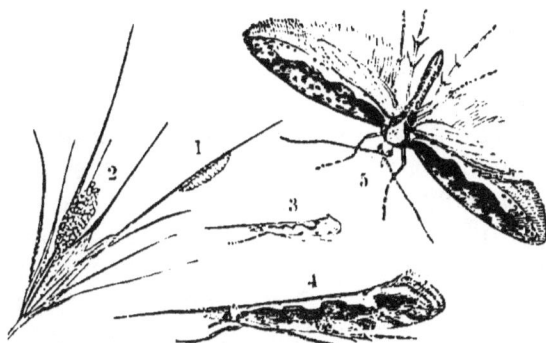

Moth (nat. size and magnified), caterpillar and cocoon.

When full fed they spin a cocoon of threads on the remains of the Turnip-leaves or on the ground, formed of such open network that the chrysalis can be seen through it. In the specimens observed this light cocoon was left open at each end, so as to allow the chrysalis to leave its old caterpillar-skin outside at one end, and the moth (on coming out of the chrysalis) to escape at the other.

The chrysalis is greyish white, with several black streaks down the back and sides. The moth hatches from it in about ten to eighteen days, and to the naked eye appears not unlike the Clothes-moth. When magnified it will be seen (as figured above) that the fore wings are long and narrow, with several pale spots on the fore edge, and a white or ochreous stripe along the hinder edge, this stripe being waved so that when the moth is at rest the two edges of the wings laid flat along the back form a row of pale diamond-shaped markings, whence the name of "Diamond-back Moth."

The hinder wings have a very long fringe. There
appear to be a succession of broods, as the moth is
observable from the end of June until October.—('Farm
Insects,' and J. O. W., in 'Gard. Chron. and Ag. Gazette.')

PREVENTION AND REMEDIES.—This caterpillar is too
small to be conveniently got rid of by hand-picking, and,
as it feeds chiefly beneath the leaves, it is thereby
protected from most attempts to injure it; so that hot
lime, soot, and salt have all been tried without more
than partial success.

Something more effective might possibly be done
(as with Turnip Sawfly) by taking a scuffler between the
rows with a bough of Fir or Broom fixed so as to sweep
the grubs off the leaves. From the habit of these
caterpillars of throwing themselves down and hanging
by their threads on being alarmed, many would be so
thoroughly swept away that they could not get back
again. In bad cases of attack, it would be worth while
to try this plan rather than have the whole crop
destroyed; or it might answer to send a man and boy
through the field, one with a bough to sweep with, the
other with soot or (in careful hands) with gas-lime, to
throw under the plants on the fallen grubs. Again, the
mixture mentioned at p. 151, of gas-lime, lime, and soot,
would be safe in all hands, do good to the plants, and be
very bad for the caterpillars.

The growth that this or any other suitable manure
would cause would be very favourable. In the severe
attack of this caterpillar in 1851, it was observed that
the "growing rains" towards the latter part of July
saved some of the crops; also that in those badly
attacked the hoed and singled portions perished, whilst
in the parts not hoed enough was saved for about half a
crop.

This caterpillar rarely occurs, but when it does is apt
—unless prompt measures are taken—to sweep the crop
fairly away.

Heart and Dart Moth. *Noctua (Agrotis) exclamationis*, Linn.

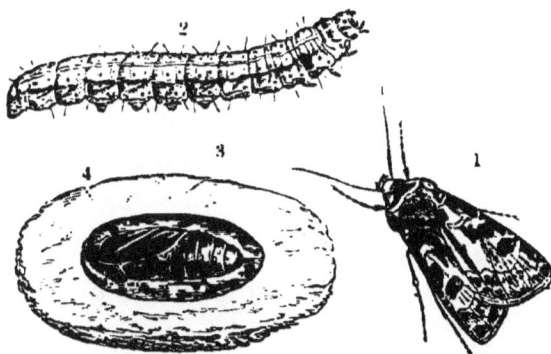

Moth, caterpillar, and chrysalis in earth-cell.

The caterpillars of this Moth live during the day just under the surface of the earth, or beneath clods, stones, loose rubbish, and such like places; and from this peculiar habit they (as well as a few other very destructive kinds) are commonly known as *"surface caterpillars."* They curl themselves up when disturbed, can walk fast, and can bury themselves in a few seconds; they come out in the evening and feed during the night, passing from one plant to another and gnawing the leaves off at the crown, so as to do far more mischief than those that only eat straightforwards at one spot.

This caterpillar is about an inch and a half long when full grown, of a dull lilac colour, with a paler stripe down the back, having one dark line along each edge, and a double one along the centre of the back. Beneath it is whitish green; the head is brown, with black jaws, and the first ring of the body behind the head is horny and brown above; all the other rings have little tubercles on them, each tubercle having a short hair growing out of it. The caterpillar is flattish and has little power of holding on with its feet, so that it readily falls when the plant is shaken. When full fed it forms a cell in the earth in which it changes to a rust-coloured chrysalis,

M

from which the moth comes out in June of the following year, or sometimes a little earlier.

The moth is of a clay-colour, with markings as at fig. 1; the spot behind the head is black; the upper wings are dark towards the front edge; the under wings in the male are white, with the upper margin and nerves brownish; in the female the under wings are dark brown.—('Farm Insects,' and 'Gard. Chron. and Ag. Gazette,' 1844.)

PREVENTION AND REMEDIES.—The moth may be seen towards the close of day, flying about neglected weedy spots, and the caterpillar may be found during the day under clods of earth, rubbish, &c., as mentioned above; therefore clearing away weeds and useless encumbrances on the ground is of great service.

The caterpillar comes out at night, and therefore hand-picking at night does good; but probably the application of waterings, or dressings, of anything offensive or injurious to the caterpillars, would be a much better plan.

Watering the infested ground with gas-water has been found effective in destroying these grubs. A thick coat of soot dug into the surface is of service; also it has been found (as with the caterpillars of the Turnip Moth) that when Cabbage and Cauliflower-plants are going off with these grubs, scattering a handful of soot round each plant and earthing it up will make it throw out fresh healthy root-fibres, and the plants will recover and do perfectly well.

This treatment is worth noting, for here, as with several other kinds of caterpillar-attack, the plant dies, *not* from being eaten up, but simply from starvation. The grub or caterpillar goes wandering from plant to plant, gnawing and injuring numbers successively so that they cannot draw up their full supply of food; and if soot or any other material is applied that drives away these creatures, and acts as a stimulant to the plant, it will often save the crop. The grubs grow quickly and require so much food that the plants fail quickly under their attack; as soon, therefore, as a plant is seen to be flagging, the roots should be examined, and, if the caterpillar is there, treatment should be applied. Guano is a very useful application in such cases, as it is good for plant-life, and often injurious to insect-life; but it should be borne in mind that, as soon as the presence of the grub is discovered, the remedy should be applied *at once.*

In this instance, as with others of the surface caterpillars, thorough cultivation of the ground, so as to clear it of weeds on which the grubs may live before a new crop is put in, and a thorough turning over of the soil, so as to bring as many as possible of the grubs to the surface where they will be picked up by birds; and, generally, to put the soil in a state favourable to the in-coming crop, is very important.

Turnip Sawfly. *Athalia spinarum*, Fabricius.

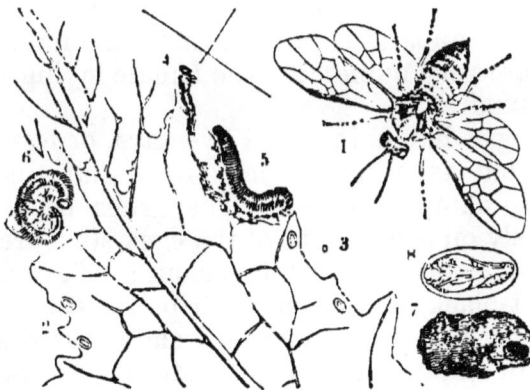

Caterpillars, pupa, and pupa-case. Sawfly, magnified, with lines showing nat. size.

The caterpillars of this Sawfly, which are known under various names, as "Blacks," "Black Palmers," "Niggers," &c., appear from time to time in very large

numbers, and do serious damage, sometimes clearing
the leafage of a whole field of Turnips, excepting such of
the veins as are too hard to be eaten, in the course of a
few days.

Taking the dates of some attacks which are especially
recorded, we find them noticed in 1756, 1760, 1782,
1806, 1818, 1833, 1835, and the following years up to
1838; in 1782 it was estimated that about two-thirds of
the Turnip-ground in Norfolk had consequently to be
ploughed and re-sown; but the worst attack recorded took
place in the dry summer of 1835, when the injury to the
Turnips extended as far north as Durham; and in the
southern counties, from Somerset to Kent, the crop was
a failure, the second and even the third sowing being
devoured by these "niggers."

The Turnip Sawflies which produce these caterpillars
are dull and torpid in moist and cloudy weather, but
they thrive and fly actively in warmth and sunshine;
and it is during the period when the sunshine is hottest
that pairing takes place.

The eggs are laid one by one in small slits cut for
them in the leaf by the saw-like ovipositor of the female
(from which apparatus the Sawflies are named); and
the eggs are very numerous, one female laying from two
to three hundred.

These eggs hatch in about five days, or less in warm
weather, but take more than twice that time if the
weather is damp and cold. The grubs begin to feed
immediately on coming out of the eggs, and are at first
usually of a greenish white colour; afterwards they
become jet-black, with a paler stripe on each side, and
when nearly full-grown they are slate-colour and pale
beneath, in addition to the pale stripe just mentioned.
Before their first change of skin (or moult) they cling
to the leaf, and, if disturbed, let themselves down by a
thread and go back again up it at pleasure; afterwards
they fall down, having no power to spin threads at this
stage of growth, and remain awhile as if dead, and then
crawl back again up the stems to the leaves.

These caterpillars have in all twenty-two feet, consisting of a pair of "true feet," horny, and furnished with claws on each of the three segments next the head; a pair of "sucker-feet" (fleshy cylindrical masses by which the caterpillar can hold fast at pleasure) on each of the succeeding eight segments, excepting the fourth from the head, and another pair of sucker-feet at the end of the tail—thus having only one segment, besides

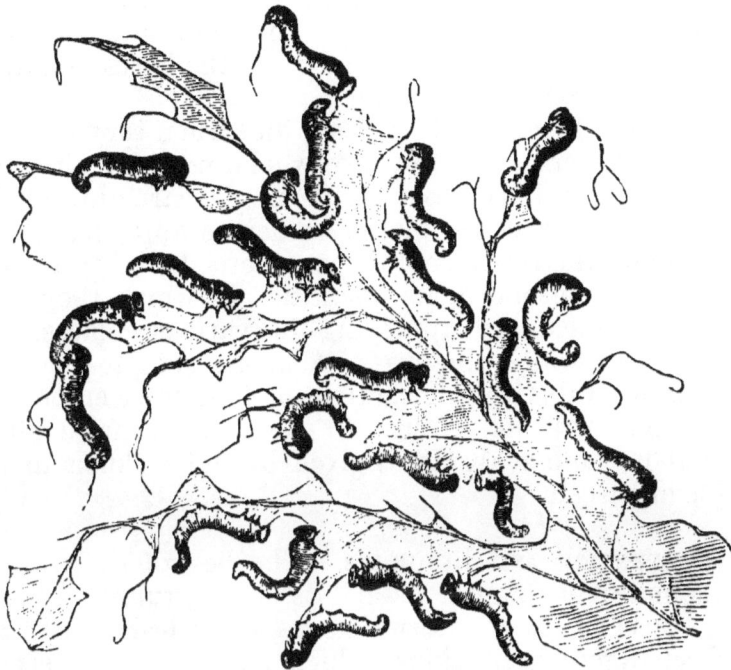

Sawfly caterpillars destroying Turnip-leaf.

the head, unfurnished with feet. They greatly enjoy being exposed to the full heat of the sun. When full grown, which is in about three weeks, they go down into the earth, spin a silken cocoon, which is smooth and white inside, but not easily distinguishable from the earth, which sticks to it externally; and from these cocoons the Sawflies come out in about three weeks in the early summer, and are ready to lay eggs and start a new attack immediately. Later in the season, three months pass before the caterpillars turn to chrysalids, and many of the autumn brood are believed to remain in the cocoons during the winter, and not to change to chrysalids (and thence to Sawflies) till the next spring.

These flies are very pretty, of a bright orange, with a deeper reddish colour just behind the black head; the four transparent wings are netted over with veins, and are yellowish towards the base; the legs are stout and short, with the shanks hairy; and the feet are whitish, with the tips of the joints and all of the lowest joint, as well as the claws on it, black.

The mischief caused by these flies when they occur in large numbers is simply overwhelming, and often (in such cases) only ceases with the total destruction of the crop, in consequence of the voracious appetites of the grubs and the rapid succession of broods. For further details of the enormous quantities in which they have been recorded as appearing in various places, the swarms in which they pass from one spot to another, and much valuable information too long to be entered on here, the reader is referred to the account (from which the above note is abridged) given by John Curtis in his ' Farm Insects.'

PREVENTION AND REMEDIES.—The best of these are to be found by looking at the habits of the grubs.

If these grubs or caterpillars are disturbed whilst they are changing their skins, which happens every six or seven days during the three weeks in which they continue feeding in grub form, they die; for if they loose hold with the pair of feet at the tip of the tail during this operation they cannot fix them again, consequently they have nothing they can pull against to drag themselves out of the old tight skin, and therefore they perish in it. Also—as mentioned above—if alarmed, they drop from the leaf, and after the first few days they have no power of spinning a thread, consequently have some difficulty in getting back again.

Looking at these points, it has been found useful to dislodge the grubs by sweeping the Turnip-leaves with boughs of light leafage or twigs, such as Fir, Furze, or Broom.

Many different ways are noted—such as fastening the boughs on a cart-rope, held by a man at each end, which thus may be dragged along so as to brush the Turnips; or fixing them to a bar supported by two wheels so that the boughs may lightly sweep the leaves as they pass over them; or again, fixing a good-sized Fir-branch or bunch of Broom in front of a scuffler, and thus, whilst the blades do their regular work, the branches sweep down the grubs, many of which are killed or do not come up again.

In slight attacks, it is of use for a man to brush the plants with a light bough, and in this case the grubs can be stamped on, or lime or soot thrown on them, and also over the leaves, which will do something towards protection.

It is also stated to have been found serviceable to drive sheep through an infested field. The passage of the sheep disturbs the caterpillars, many of which fall from the leaves and are trodden under foot; but the desirableness of the remedy is somewhat doubtful (especially on heavy land), excepting in very extreme cases of attack.

When the attack is bad the Turnips should not on any account be hoed or thinned (the grubs will thin them only too much), and, if left alone, enough may escape for a crop; also the hoeing does not hurt the Sawflies or their grubs to any useful degree, and the pests from the destroyed plants will move on so as to be in double force on the remaining part of the crop.

Thick-sowing is of advantage against Sawfly, as well as against the common Turnip Fly, both for the reason just given of some part of the crop escaping, and also because the Sawfly likes sunshine and dryness. Where the leaves are plentiful there is more moisture and shade, and these parts are safer from attack; and also

in the hot dry seasons in which these Sawflies thrive the
thicker vegetation will help on the plant, which loves
moisture.

For these reasons—that is, a plentiful supply of
moisture being needed for the proper growth of the
Turnip-plant, and moisture being prejudicial to the
Sawfly—thorough watering with liquid manure by means
of the water-cart on a dull day, or after sunset, would be
most beneficial. It would stimulate the growth of the
plant, and destroy many of the caterpillars by washing
them off the leaves into the wet soil. A dull day for the
watering, or a time when the sun is not shining, is
desirable) to avoid a sudden chill, which would be bad for
the leafage), or the sunshine burning spots where the
water might have run together in drops, on the leaves.

In all attacks on leafage it is of the utmost importance
to keep up the strength of the plant; the damage done
is by the grubs destroying what for its own vegetable
uses are in fact its breathing and digestive organs. It
cannot live without its leaves, and if a larger portion of
these are eaten off each day than is replaced by growth,
it fails just in that proportion. By the application in
moist weather of nitrogenous or ammoniacal manure, or
by watering with liquid manure in drought, a plentiful
supply of food is brought to bear at once, and thus we
may possibly tide the plants over the difficulty and save
the crop within the bounds of remunerative outlay.—
('Notes of Observations of Injurious Insects,' 'Farm
Insects,' &c.)

Turnip Gall Weevil. *Ceutorhynchus sulcicollis*, Stephens.

This Weevil is injurious to Swede and White Turnips
by causing the growth of knobs or galls on the surface of
the bulbs, similar to those that it causes on the roots of
Cabbage.

On opening these knobs one or more maggots will be
found inside, of the same species as those in the Cabbage,

but commonly of a yellowish colour when they are feeding inside galls of the Swede Turnip. As the habits of this Weevil in its various stages, and also its power of

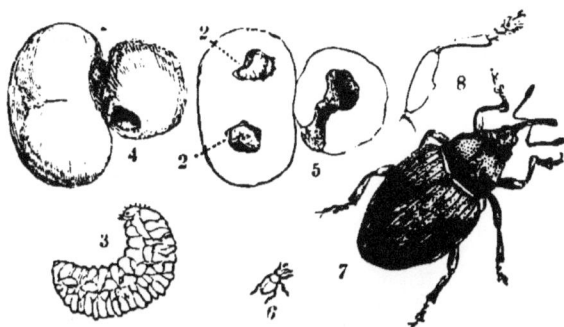

1—5, Gall with maggots nat. size and magnified; 6 and 7, Weevil, nat. size and magnified; 8, leg of Weevil, magnified.

endurance of cold in the maggot state, appear to be the same whether it feeds on Cabbage or Turnip, the reader is referred for the life-history to the notes of Cabbage Gall Weevil, p. 41.

PREVENTION AND REMEDIES.—These galls are rarely numerous enough to injure the Turnip, and commonly the soft juicy swellings do not lessen the quantity of nutritive material in the bulb to an important amount; still they are better away.

Clean cultivation and regular rotation of crops is a means of keeping the attacks of the weevil in check, and good dressings of chalk and lime are excellent preventives; also gas-lime, sown broadcast over the soil intended for Cabbage and then pointed-in, is a good means of getting clear of the maggot; see also pp. 43, 44.

PART II.

FOREST TREES

AND

THE INSECTS THAT INJURE THEM.

FOREST TREES

AND

THE INSECTS THAT INJURE THEM.

———◆———

ASH.

Ash-bark Beetle. *Hylesinus fraxini*, Fab.

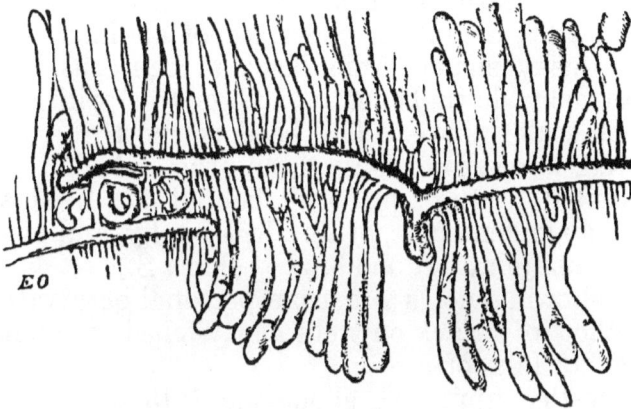

Workings of *Hylesinus fraxini*, showing forked "mother gallery,"
with larval galleries from the sides.

THE *Hylesinus fraxini* is injurious, both in the beetle
and grub state, to Ash trees, by boring galleries beneath
the bark, sometimes slightly cutting into the outside
wood of the tree. The beetles are often attracted in

large numbers by newly-felled Ash trunks, in the bark of which they propagate, and from whence the new brood spreads to the neighbouring trees, mainly attacking those that are sickly or decayed; or young trees, which they sometimes injure to a serious extent. The damage is caused not only by the bark being loosened and the regular circulation of the sap interfered with, but also by the multitude of small holes which the beetles bore in escaping after they are developed, allowing rain or moisture to soak into the substance of the bark and cause decay.

The larvæ are small whitish fleshy legless maggots, much like those of *Scolytus;* the head is furnished with a pair of jaws by means of which the maggot gnaws its gallery beneath the bark.

The beetles are about the sixth of an inch long, of various dusky shades from black to ochreous, covered with an ashy down beneath, and mottled with ashy or brownish scales above. The head is short and robust, horns red, lowest joint longest, and the end club-shaped and pointed at the tip; body behind the head stout, convex; abdomen short, ovate; legs pitchy, and feet red, with the last joint but one bifid.—(J. F. S.)

The following notes are from personal observation of the method of attack on trees newly-felled in the neighbourhood of Isleworth :—

The beetles appeared about the 19th of April, and after wandering about on the bark for a few days the workings were begun by each beetle boring a circular hole just large enough to admit it. Here it was shortly joined by a companion, and pairing took place.

At about half an inch at most from the entrance, instead of carrying the tunnel straight forward (as with those of the Elm-bark Beetle), the workings forked, and the two galleries were carried on to right and left, until, in about five weeks they were at their full length, and the working was shaped much like a T with a short stem. During this time one beetle was usually to be found in each of the side galleries, but occasionally they were together.

By the 4th of July most of the parent beetles were
dead in their burrows, and a few of the grubs hatched
from the eggs which had been laid along each side of the
tunnels had begun their borings; about three weeks
later these larval tunnels were to be found completed,
and pupæ were then fairly numerous in the cells formed
by each larva at the end of its gallery. The beetles
began to appear about the 10th of August: each beetle
as it developed eating its way out, and soon, from the
number of these perforations, giving the bark an appearance as if it had been riddled by shot-holes.—('Entomologist.')

PREVENTION AND REMEDIES.—The damage caused by
these beetles is chiefly to decayed or sickly trees, or to
young trees; the attacks on felled trunks are only of
importance by serving to propagate the pest.

Attention to suitable locality and soil, and such
management as may keep the trees in health, is the
best method of prevention. The Ash has a large
number of lateral fibrous roots, and likes "a good dry
soil within reach of water." "A free loam with a mixture
of gravel" is considered suitable, but a boggy soil or low
swampy ground, or stiff clay, are not suitable to its
continuous healthy growth.

Judicious thinning and removal of injured or infested
branches are important matters.

The Ash likes shelter, but if plantations are allowed
to run on too long without thinning, it suffers much
from the sudden exposure; and where dead or dying
boughs have not been removed, these attract insect-
attack, which spreads till the ruin of the whole tree
ensues.

Careful removal of dead, or decaying, or sickly boughs,
or such as are suffering from insect-attack, is highly
desirable.

Where felled wood is found to be attracting attack
(which may be easily known towards the end of April by
the little heap of chips lying at the mouth or below the

mouth of each beetle-burrow) the removal of the bark is a sure remedy. If, however, barking is a heavier operation than is wished, a good thick coat of mud laid on the timber and well rubbed into all the crannies is a very fair protection, particularly if some paraffin is stirred into the mud before application.

With regard to attack on live trees, this should be watched for between the middle of April and of May, and if chips are found to be thrown out from small burrows about as large as a shot-hole, measures should be immediately taken.

The best method is probably to set a man with ladder and pail, to rub a good coat of soft-soap into accessible parts of the tree by means of a common scrubbing-brush, or in any other way that may be more convenient.

In a case like this, where the season of attack probably only lasts for a short time and the injury is often to a few trees, it is well worth while to stop it at once; and a coating of any substance that is offensive to the insect, and, like soft-soap, chokes up its breathing-pores, clings to its limbs, and fills up its boring, is very serviceable.

Any application which is not injurious to the tree, and will gradually be washed off by the rain, will be of use; and probably (as these Bark-feeding Beetles do not frequent dung) a thorough coating of cow-dung mixed with clay and water, and laid on thickly with a long-handled whitewasher's brush, would answer well.

Ash-bark Scale. *Chionaspis fraxini.*

This Scale-insect is sometimes to be found in large numbers on the bark of Ash trees, infesting the parts which are still soft enough for it to pierce with its sucker, by which means it injures the tree both by drawing away the sap and also by the innumerable punctures it makes into the tissues.

The Scales containing the female and her eggs are

white, of a soft papery consistency, and from their soft texture and the large numbers fixed side by side or partly on each other are often compressed into all kinds of irregular shapes. When perfect they are somewhat mussel-shaped, but rather broader for their length, and without much difference in the shape of the two sides.

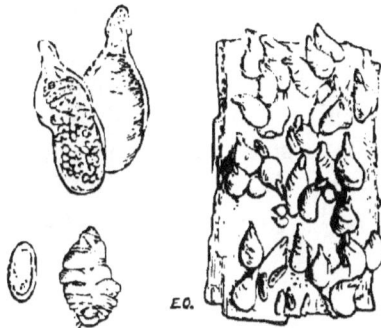

Scale, containing female and eggs; egg, showing larva within; female, much shrivelled; all much magnified. Scales on bark, slightly magnified.

On raising this Scale during the winter months the shrivelled body of the female will be found beneath, towards the smallest extremity, the rest of the space being filled with a multitude of crimson eggs containing the young Scales, which may be seen with a high magnifying power, as figured above, through the transparent membrane of the egg. The mother-insect in my specimens was too much shrivelled for the parts to be clearly distinguishable, but appeared (as usual with female Scale-insects) to be a mere fleshy mass, with a sucker so far beneath it as to seem to arise from the breast, and without legs or wings. Amongst the mussel-shaped Scales were others only about a third of their length, and with parallel sides, which, from comparison with other species, I presume to be the pupa of the winged male; but my observations being taken in winter, and this Ash-bark Scale being rarely studied, I am not able to give its complete life-history.

N

PREVENTION AND REMEDIES.—Where this Scale occurs on trees in nurseries it is desirable to clear it, which may be done by the use of any of the applications mentioned under the head of "Apple Scale," such as brushings with soft-soap, &c.; or if only a few trees are attacked it is well to cut them down and burn them, to keep the attack from spreading, as, though not material at first, in time it affects the health of the tree.

Little information is given as to causes disposing to attack, but in the case of specimens forwarded from West Gloucestershire, the infested Ash grew in a wet, tenacious, lias clay, sometimes completely sodden with moisture, and precisely the situation considered to be prejudicial to healthy growth of the tree.—(J. U.)

BIRCH.

Bud Gall-Mite. *Phytoptus* (? sp.)

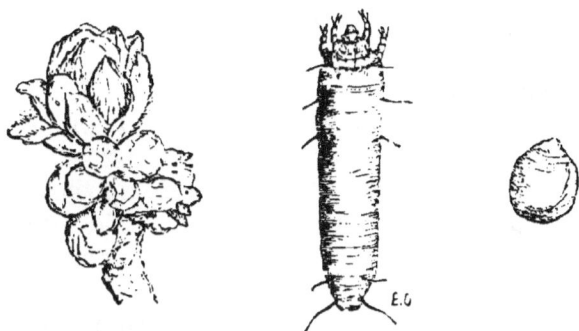

Infested buds, magnified; Gall-mite and egg, much magnified.

The diseased growths formed of irregular masses of twigs sometimes nearly a yard long, which are often seen hanging from Birch-boughs, and are commonly known as "Witch Knots," or "Witches' Brooms," are caused by this Gall-mite.

Gall-mites, or *Phytoptidæ*, are a subfamily of the order *Arachnoidea ;* of quite a distinct class from insects, and easily distinguishable from them (in any stage of insect-life) by having two pairs of legs; but as they are the cause of a good deal of injury to plant-life, it seems desirable just to mention them.

The Gall-mite of the Birch is so minute as to be imperceptible to the naked eye, except when collected in masses; when magnified it appears as figured above ; dull white, long and cylindrical in shape, much wrinkled transversely, and furnished with two pairs of legs placed together near the foremost extremity, which can hardly be called the head, but contains the feeding apparatus by which the various species of Gall-mites cause much injury to the soft surface of young leaves or other part of their food-plants.

The eggs may be found at various seasons during the whole year, for I have found them with the Gall-mites hatching from them amongst diseased bud-scales in February, and at the end of August eggs were still to be found.

The formation of the Witch-knots begins with a diseased growth of the Mite-infested bud, which is distinguishable by its swelled, irregular loosely-opened appearance, from the smooth and pointed shape of the buds in healthy condition; in the next stage (also in figure) the attacked shoot is thickly covered by the buds, which in healthy growth would have been distributed at distances of some inches along it. As time goes on, repeated forkings of the twigs from these unhealthy and infested buds, and from successive growths of the same kind, give rise to the knotted and confused masses known as Witches' Brooms. Some-times these make little progress, and the knot merely resembles a rough mass like an old rook's-nest thrown down and hanging loosely from the Birch-bough; sometimes the twigs regain healthy growth, and pushing on for as much as a yard in length form a pendant mass of some beauty, from the delicacy and gracefulness of the sprays.

The formation of the Witch-knots was first traced to the action of *Phytopti* in England in 1877 (see papers in 'Entomologist' and in 'Gardeners' Chronicle' for that year); and it was stated by Mr. A. Murray, in his comments on the observations, that the *Phytoptus* above described which infests the buds is quite distinct from the species producing the appearance on the back of the leaves known by Schumacher as *Erineum betulinum*, and as far as I am aware it has not yet received a specific name. It is of considerable interest as producing a diseased growth of the twigs, whereas most of the *Phytopti* simply feed on or produce diseased growths of the soft tissues of the leaves.—('Entomologist, 'and 'Gard. Chron.')

PREVENTION AND REMEDIES.—Witches' Brooms should be cut off and burnt, and in cases where the tree is much infested with the small gradually-forming tufts of diseased growth it is desirable to cut it down and burn the Mite-infested twigs. The Gall-mites have no power of flying, but the wind wafts them about on leaves or broken twigs, or birds carry them in their plumage, and when once well established the attack spreads regularly onwards (as may be seen on the roadside trees at Spring Grove, near Isleworth), slowly but steadily to the neighbouring Birches.

Note.—The Black Currant and the Filbert are sometimes injured by Gall-mites, known respectively as *Phytoptus ribis* and *Calycophthora avellanœ*. The presence of these may be known by the swollen and partly-opened condition of the bud, like that figured at the end of the shoot. In these cases it would be desirable, if possible, to clear away the bushes or the boughs that are infested; probably all the various remedies or means of prevention for attack of Red Spider, such as dressings of quick-lime or gas-lime beneath the bushes, stirring the surface soil, and measures of good cultivation generally, would be of service. In the worst instances that I have seen of Gall-mite infested buds on Filberts, the bushes appeared to have been neglected for several years, and were also overshadowed by trees.

ELM.

Elm-bark Beetle. *Scolytus destructor*, Oliv.

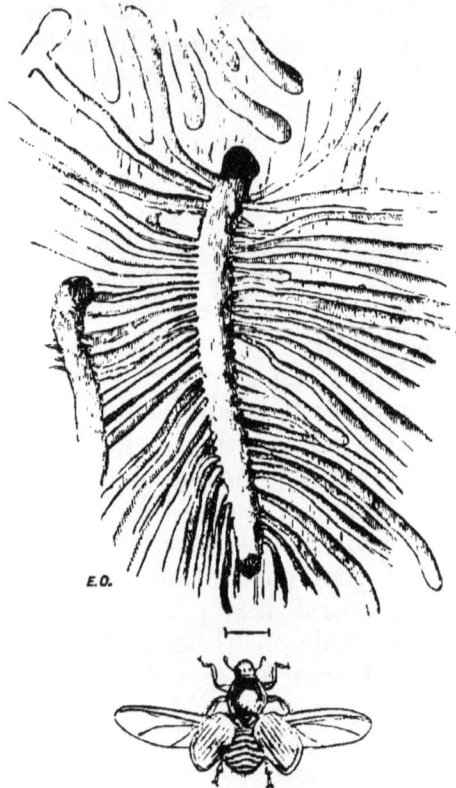

E.O.

Beetle, magnified; and beetle and maggot workings in Elm bark.

This beetle is well known as causing much injury to
Elm trees by means of the galleries that it bores between
the bark and the wood, mainly in the soft inner bark,
but so as also to leave just a slight trace of the working
on the surface of the wood.

The females may be seen early in June, making their
preparations for egg-laying by working their way along
the bottom of cracks in the bark, which they widen for

some distance before beginning to burrow, so that the real opening of the galleries may be at some distance from the heap of rejected matter or little heap of wood-dust that marks the first point of entrance.

The male is present for only a short time after the burrow is begun, before egg-laying commences.

The burrow of the parent beetle is usually about three to five inches long, and takes about three weeks to form. The eggs are laid along each side of it, and are a hundred and upwards in number. As many as a hundred and sixty have been observed.

The young grubs, when hatched, start at right angles from the parent gallery, and gnaw their way onwards, the burrows gradually increasing in size and curving to allow room for the growing size of the tenant (as shown in the fig.)

Most of the larvæ are full fed towards the end of July, when some turn to pupæ at the end of their burrows, and the beetles from these pierce the bark and come out from the tree during August. The greater number, however, of the grubs, appear to form a little chamber either just within the wood or in the thick bark, where they pass the winter, and come out as beetles about the end of May; thus, in case of the bark being removed or falling from the tree, although the beetle-maggots in the open galleries beneath it are exposed, to be cleared off by the birds, the others buried in their chambers, with the openings stopped up by the rejected matter, are safe from attack.

The maggot is whitish, curved, tapering bluntly to the tail, fleshy, much wrinkled across, and legless.

The beetles are black, 1½ to 3 lines long, with rounded rough head and reddish horns; wing-cases rounded at the sides, and cut short at the tip, pitted lengthwise with rows of dots, with irregular punctures between, glossy, and sometimes of a pitchy colour or rusty red; abdomen cut short, slantwise beneath; legs pitchy; feet reddish.

It has been observed that these beetles prefer a tree that has already been attacked rather than a young and vigorous one, and it is easy to tell where they are or have been present by the great number of small holes, as if the bark had been pierced by shot or holes made by a bradawl, and also (whilst the beetles are boring their way out) by the wood-dust observable on the bark, or fallen on the ground beneath the openings of their burrows.

The circumstance of *Scolytus* attack, and sickly growth of the tree or decay of the bark occurring together, has given rise to much discussion as to whether the *Scolytus* attack caused the decay, or weakened health induced attack.

It is observed by Dr. Chapman, from whom I take much of the above life-history (see 'Entomologist's Monthly Magazine,' 1869, pp. 126, 127), that healthy growing trees are supposed to repel the attacks of this genus of beetles by pouring sap into their burrows. He notes that in the case of the *Scolytus pruni* he had observed "burrows less than one inch long, some of which, containing a few eggs already laid, had been abandoned uncompleted by the beetles, apparently on account of the presence of a fluid which must have been sap, as no rain had fallen to account for it."—(Ent. Mo. Mag.; Illus. Brit. Ent.; &c.)

PREVENTION AND REMEDIES. — The best method of remedy appears to be that adopted with great success in France by M. Robert, after careful observation of the circumstances which stopped the operations of the female beetle when gnawing her gallery for egg-laying, or which disagreed with or destroyed the maggots, and is based in part on similar observations of the effect of flow of sap to those noticed in England by Dr. Chapman.

It appeared, on examination, that the grubs died if they were not well protected from the drying action of the air; on the other hand, if there was a very large amount of sap in the vegetable tissues that they fed on, this also killed them, and it was observed that, when the female was boring through the bark, if a flow of

sap took place she abandoned the spot and went elsewhere. It was also noticed that the attack (that is, the
boring of the galleries which separates much of the
bark from the wood), is usually under thick old bark,
such as that of old Elm trunks, rather than under the
thinner bark of the branches. Working on these observations, M. Robert had strips of about two inches wide
cut out of the bark from the large boughs down the
trunk to the ground, and it was found that where the
young bark pressed forward to heal the wound, and a
vigorous flow of sap took place, that many of the maggots
near it were killed, the bark which had not been entirely
undermined was consolidated, and the health of the tree
was improved.

Working on from this M. Robert tried the more
extended treatment of paring off the outer bark, a
practice much used in Normandy and sometimes in
England for restoring vigour of growth to bark-bound
Apple trees, and noted by Andrew Knight as giving
a great stimulus to vegetation. M. Robert had the whole
of the rough outer bark removed from the Elm (this
may be done conveniently by a scraping-knife shaped
like a spokeshave). This operation caused a great flow
of sap in the inner lining of the bark (the liber), and
the grubs of the *Scolytus* beetle were found in almost all
cases to perish shortly after. Whether this occurred
from the altered sap disagreeing with them, or from the
greater amount of moisture round them, or from the
maggots being more exposed to atmospheric changes, or
any other cause, was not ascertained, but the trees that
were experimented on were cleared of the maggots. The
treatment was applied on a large scale, and the barked
trees were found, after examination by the Commissioners
of the Institute at two different periods, to be in more
vigorous health than the neighbouring ones of which the
bark was untouched. More than two thousand Elms
were thus treated.

This account is abridged from the leading article in
the ' Gardener's Chronicle and Agricultural Gazette ' for

April 29, 1848, and the method is well worth trying in our public and private parks. It is not expensive; the principle on which it acts as regards vegetable growth is a well known one, and as regards insect health it is also well known that a sudden flow of the sap that they feed on, or a sudden increase of moisture round them, is very productive of unhealthiness or of fatal diarrhœa to vegetable-feeding grubs.

A somewhat similar process was tried by the Botanic Society in 1842 on trees infested by the *Scolytus destructor* in the belt of Elms encircling their garden in the Regent's Park, London; "it consists in divesting the tree of its rough outer bark, being careful at the infested parts to go deep enough to destroy the young larvæ, and dressing with the usual mixture of lime and cow-dung." This operation was found very successful, and details with illustrations were given in a paper read in 1848 before the Botanic Society.

Various applications have been recommended, such as brushing the bark of infested trees with coal-tar or with whitewash, in order to keep off the beetle attack. Anything of this kind that would make the surface unpleasant to the beetle would certainly be of use so long as it was not of a nature to hurt the tree, and if previously the very rugged bark was partially smoothed it would make the application of whatever mixture might be chosen easier and more thorough.

Anything that would catch the beetles, either going into or out from the bark, like coal-tar, would be particularly useful, and probably strong-smelling and greasy mixtures, such as fish-oil soft-soap, would do much good.

Washing down the trunks of attacked trees has not been suggested, but, looking at the dislike of the female beetle to moisture in her burrow, it would be worth while, in the case of single trees which it was an object to preserve, to drench the bark daily from a garden-engine for a short time when the beetles were seen (or known by the wood-dust thrown out) to be at work forming burrows for egg-laying.

The possibility of carrying out the important point of clearing away or treating infested standing trees depends, of course, on local circumstances; but, whatever care is exercised in other ways, it is very unlikely that much good will be done in lessening attack so long as the inexcusable practice continues of leaving the felled trunks of infested Elms lying, *with their bark still on*, when containing myriads of these maggots, which are all getting ready shortly to change to perfect beetles, and to fly to the nearest growing Elms. Such neglected trunks may be seen in our parks and rural wood-yards all over the country, where, without difficulty, the hand may be run under the bark so as to detach feet and yards in length from the trunk all swarming with white *Scolytus* maggots in their narrow galleries.

This bark, with its contents, ought never to be permitted to remain. Where it is loose it may be cleared of many of the maggots by stripping it off and letting the poultry have access to it; or, if still partly adhering, it may be ripped from the wood by barking tools, and burnt, but it is a tangible and serious cause of injury; and if our landed proprietors were fully aware of the mischief thus caused to their own trees and those of the neighbourhood they would quickly get rid of it.—('Gard. Chron. and Ag. Gazette,' Ed., &c.)

Goat Moth. *Cossus ligniperda*, Fab.

The caterpillars of this Moth are injurious to many kinds of our timber and fruit trees, as Elm, Ash, Oak, Beech, Lime, Willow, Poplar, Apple, Walnut, &c., by gnawing large galleries in the live wood.

The Moth lays her eggs in crevices in the bark commonly at the lowest part of the tree, and the caterpillars which hatch from these eggs feed at first in the bark, or between the bark and the wood; as they grow stronger they eat their way into the wood, and form chambers and

galleries of various size and width, some as large as a
man's finger.

The caterpillar is three inches long (or mroe), of a yellow
colour with a black head, two black spots on the ring
behind the head, and either a row of dark reddish
patches or a stripe of the same colour along the back;
when young it is of a more fleshy colour. It has the
power of exuding an oily fluid from its mouth with a
remarkably pungent goat-like smell, whence the name of
the moth.   Infested trees may often be known by this
disagreeable smell, and sometimes by heaps of dirt or
wood-dust thrown out by the caterpillars lying below the
entrance to their burrows.   Sometimes the workings of
the caterpillar are shown by a dark wet patch formed
just below the entrance by the sap oozing from within.

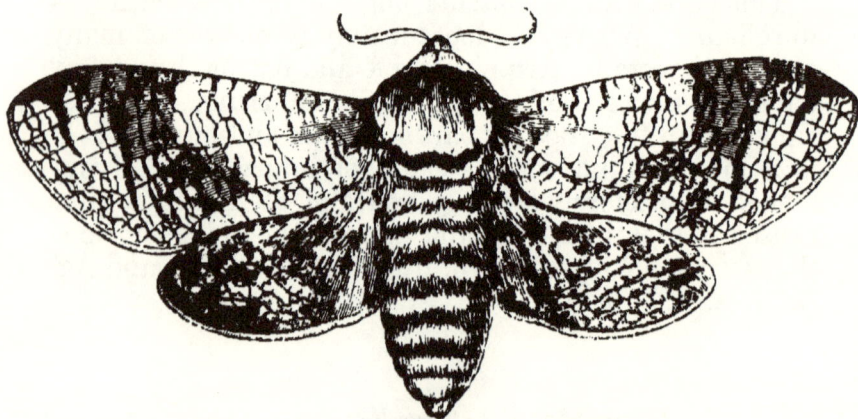

Goat Moth.

During the winter they lie quiet, otherwise they feed
for a period of three years, and, when ready to change,
form cocoons of little bits of wood roughly spun together
just inside the entrance of their burrows, in which they
turn to a reddish brown chrysalis.   Shortly before the
moth is ready to emerge the chrysalis forces itself partly
through the cocoon, where the empty case remains
sticking out from the tree, and is a useful guide as to
infested timber.

The moth is upwards of three inches in the spread of the fore wings, which are mottled with ashy white, and rich brown with many irregular black streaks and markings ; the hinder wings are of a more dingy colour, with the markings less distinct ; the head dusky brown ; body

Young caterpillar, and chrysalis.

between the wings marked across with dark brown and grey or ochreous ; the abdomen brown and grey in alternate rings. It is to be seen at the end of June and beginning of July.—(Loudon's 'Arboretum,' 'Brit. Moths,' &c.)

PREVENTION AND REMEDIES. — The moths are heavy and sluggish, and may be taken easily by hand as they rest quietly during the day on the bark of the tree out of which they hatched.

The caterpillars sometimes leave the trees, and may be found straying about in May and in the autumn, and in such case they should always be destroyed ; but generally (as above mentioned) they change to chrysalids at the entrance of their burrows, and where trees are known to be infested these reddish chrysalids should be looked for during June or early in July.

Any mixture that can be laid on the tree, so as to prevent the moth laying her eggs on the bark, is useful, and a thick coating of clay and cow-dung has been found to answer well. It is a point worth consideration that the conditions needed for the existence of the live timber-feeding and the dung-feeding caterpillars are so totally different, that the instinct of the mother insect will almost certainly keep her from depositing eggs on unsuitable material; consequently, in the case of the Goat Moth and in many other kinds of insects producing wood-boring caterpillars, a good coating of cow-dung, mixed with enough clay to make it tenacious, would be a very useful application. Whale-oil (or fish-oil) soft-soap, is also stated to be a good protection, used as follows :— Several pounds of the soft-soap are mixed in a pail with warm water to about the consistency of thick paint ; the operator, who is also supplied with a bag of sand and a coarse cloth, dips the cloth in the soap and sand and rubs the bark thoroughly, and then, with a painter's brush, lays on a thick coat of oil.

This treatment is a good means of preventing oviposition, and also of rubbing off or destroying eggs that may have been laid on the bark ; but the application of oil afterwards is less desirable, relatively to the chance of hurting the tree. If only applied to old thick bark it probably would do no harm, but, where the outside of the bark had still any life in it, it would be injured by the pores being choked up. The same difficulty occurs in the use of tar ; it is often a most serviceable means of prevention when used in moderate quantity, or only on old bark, but if applied over too large a surface, or on young bark, or again, if exposure to hot sunshine should melt the tar, and so allow it to sodden into the bark, it causes much injury.

The caterpillars may be diminished in number by crushing them in their holes with a thick strong wire ; a glance at the state of the end of the wire when it is withdrawn from the hole will show whether the caterpillar has been reached or not. If the direction of the hole admits of the caterpillar being dragged out by a finer wire doubled at the end, so as to form a kind of hook, this plan is also serviceable.

Paraffin injected by a sharp-nozzled syringe with as much force as possible into the holes where the caterpillars are working is a good remedy (M. D.) ; and any fluid poisonous to the caterpillar, or which would make the wood of its hole poisonous or distasteful to it for food, would be serviceable, as tobacco-water, or a solution of soft-soap. The fluid might also be easily injected by means of a gutta-percha tube, of which one end was fitted on the sharp nozzle of a syringe and the other passed a little way up the hole.

The fumes of sulphur blown into the hole are very effective in destroying the caterpillars of the Leopard Moth (M. D.) ; and probably this application, or a strong fumigation of tobacco, would be equally serviceable in the case of the Goat Moth caterpillars.

Where a tree is much infested, it is the best plan to cut it down, split it, and destroy the caterpillars within ; as many as sixty or more caterpillars may be taken from one tree, and when in this state it will never thoroughly recover, and it becomes a centre to attract further attack, as well as one to spread infection.

" The green woodpecker preys on these caterpillars, and its stomach, on dissection, has an intolerable stench."—(J. O. W.)

LARCH.

Larch Aphis.
  ,,   "Bug."   } *Chermes laricis*, Hartig.
  ,,   "Blight."

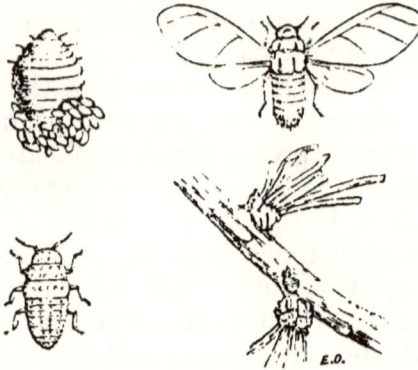

Female, with eggs, winged specimen, and larva; all magnified. Twig, with females and eggs, slightly magnified.

The attack of this Aphis, known also as Larch Chermes, Larch Bug, or Larch Blight, causes injury by means of the insects in all their stages piercing the tender bark or leaves of the Larch with their suckers, and drawing away the sap. It occurs on old as well as young trees, but is most injurious to the latter by reason of the larger proportion of the tree liable to attack.

The *Chermes laricis* never produces living young; it propagates entirely by eggs, and when the Larch-leaves are beginning to appear in the spring, the mother *Chermes* may be seen at the base of the leaf-knots along the Larch twigs, laying the eggs which will give rise to the successive generations of the year. These eggs are oval, and furnished with a kind of hair-like stalk; of a yellow or yellowish purple colour at first, which deepens in tint towards hatching-time to a dark violet. They

are laid slowly) sometimes at the rate of about five a day),
and more or less covered up with a kind of powdery down
removed off herself by the mother, and gradually are piled
round and over her till she is half-buried in them, and in
hardening drops of turpentine which she constantly
exudes with a kind of pumping motion.

This female, the mother of the colony, is of the shape
figured above, greatly magnified; wingless, with short
legs, and a strong sucker; dusky violet in colour,
becoming darker with age, and more or less covered
with a white powdery or cottony secretion. The legs
and sucker are dark, or black.

The young soon hatch; eggs may be found in the
south of England in course of laying on the 22nd of
April, and twigs swarming with young at the beginning
of May. These are of the shape figured above, with
distinctly-formed head and horns; trunk (or thorax)
with six legs; and abdomen: at first they are of a
powdery black, or violet, with several rows of tubercles
along the abdomen, and (though not showing as clearly)
also along the trunk; afterwards they change "to an
olive-yellow or clear olive-green, with horns, legs, and
sucker darker olive-green or olive-brown."—(C. L. K.)
These disperse themselves over the leaves, and, piercing
into them with their suckers, begin the work of mischief,
and the infested shoots may be known by the *Chermes*
scattered over the leafage, like little black or darkish
specks bearing bunches of white down. Later on—about
the middle of May—fully-developed winged as well as
wingless specimens may be seen: the winged females of
the shape figured, of a yellowish tint, with brown head
and horns, and various brown markings; and with wing-
veins of a yellowish green. The reader is requested to
notice that the long vein forked at the end, placed at the
fore edge of the upper wing, has only *two* side veins
from it: this veining of the wings is characteristic of the
tribe *Chermisinœ*, and distinguishes it from the three
other tribes of the *Aphididœ* (for details of wings, see
"Aphides" in Index).

The Chermes-attack continues, or may continue, unless checked by weather or special circumstances, till August or later, and the last laid eggs of the year produce again the large shapeless "Mother Chermes," the foundress of the family of each successive year, which lives through the winter, and in spring lays her eggs as above described.   Descriptions of the male *Chermes* have been given, but our best authorities on the subject consider that at present it has not been observed.— ('Mon of Brit. Aphides,' 'Die Pflanzenlause,' and observations by Ed.)

PREVENTION AND REMEDIES.—The following remedies have been of service in checking attack of Larch Bug when already commenced :—

A note is given of a plot of young Larches planted in nursery-ground a year previously, which became so badly infested with "bug" in May that they appeared as if covered with mould, with the sap exuding over the stems, so that the shoots were soft and supple, and the plants becoming rapidly exhausted.   These were watered over head with dilute paraffin, in the proportion of a wine-glassful of paraffin to a watering-can full of water, and the first application checked the depredations of the Bug.   The waterings were repeated at intervals of three or four days, for about three weeks, when the plants were entirely cleared of the Bug, and assumed a healthy and vigorous appearance.   The application was found similarly serviceable in clearing Pine Bug, and in no way injurious to the trees when applied judiciously.— (J. K.)

The following remedies have proved efficacious in destroying "Bug," and preventing attack on Larch and Silver Fir.

One method is as follows :—To every thirty-six gallons of water add half a pound of perchloride of mercury ; with this the infested trees are drenched in the early summer, when the sap is flowing freely ; a dry day is preferred for the operation, as it gives time for the solution to soak thoroughly into the bark. This has been applied to ornamental trees and plants in the nursery, and it is noted that trees operated on in 1873 continued, at the time of writing (1880), free from the "bug" and in thriving condition. This application requires to be in careful hands, *being poisonous;* Wood-peckers that fed on the poisoned insects were destroyed by it; and especial caution is given against using it to fruit-trees.—(D. F. M'K.)

Another method found serviceable was the use of lime-water prepared and applied thus:—One hundred-weight of best lime-shell to eighty gallons of clear water: slake the shells in the water, and allow it to stand for a week; drain off the clear liquid, and wash or syringe the infested trees. This was found to clear the tree thoroughly of the Bug and eggs. The trees appeared to be a little sickened for a time, but all recovered.

Washing with lime and water was also found to answer both as a remedy and means of prevention, but made the tree unsightly.—(D. F. M'K.)

The use of quick-lime, in a plantation of Larches from eight to twelve feet high, is noted as "very disagreeable, and only partially successful." Tobacco-liquor is also mentioned as being applied for Aphis-attack to Silver Firs, the solution being rubbed on the tree and branches; this was more successful than the lime, but more expensive and difficult of application.—(J. M'L.)

Looking at the good effects both of tobacco and of soft-soap for general use in clearing off Aphides, it is probable that some of the Hop-washes in which these are combined (and which might be easily applied by a garden-engine) would be very serviceable.

The following recipe from amongst those given under the head "Hops" is simple, and found reliable for regular use in the Hop-gardens:—To thirty-six gallons of water in a copper add sixty pounds of soft-soap, then add either fourteen pounds of bitter aloes or two pounds of tobacco, and boil together. For use add thirty-six gallons of water to every gallon of this liquid.—(J. W.)

The common fish-oil soft-soap manufactured by Messrs. Gibbs for use in the Hop-grounds may be purchased at ten to twelve shillings per firkin of sixty pounds.

Paris green (arsenite of copper) is a more expensive application, being sixpence per pound, but it might be worth trying mixed with dry lime. The proportion used for dressing Potatoes is one part of the green to twenty of lime; it should, however, not be entrusted to any but thoroughly careful hands, being a *deadly poison*.

In dealing with the Larch Bug we have advantage, from its flocky coat, as washings and dressings, especially those of a sticky nature, clog the down, and thus take good effect.

The amount of the presence of this *Chermes* (like that of other Aphides) appears to depend partly on such states of the weather and local atmospheric surroundings of the trees as may be suitable for increase of the insect, partly on the health of the trees, and also on their neighbourhood, to such as are infested.

Late frosts are noted as being injurious to the Larch, and favourable to increase of the "Bug"; this, presumably, for the same reasons as in other cases of Aphis-attack, that late frosts commonly accompany clear skies, with bright sunshine by day, and the sudden alternations of heat and cold are unsuitable for healthy growth, but cause a condition of sap suitable to the Aphides. In 1880, in which year the "Blight" was very prevalent, it was observed in connection with frost in June, and with hard dry winds; and looking back to published records of former years, it is noticed as "most prevalent when the frosts were very severe late in the season."

The health of the Larch depends greatly on local conditions; it suffers from drought and from exposure of its roots to sunshine, and also from a stagnant wet soil. Although it requires a constant supply of moisture for its roots, this moisture must be fresh, and free; and it needs a clear dry atmosphere, with great amount of sunshine for its leaves.

A position amongst broken rock, with plenty of good loam, either on the side of a ravine or so placed that water may constantly trickle by, may be considered the type of what is most suitable for its growth; in its natural habitats it thrives best on declivities connected with summits of perpetual snow, by the thawing of which the plants are fed, and where their heads are well exposed to sunlight.

It has been pointed out by Prof. DeCandolle that the fine slender Larch leaves having less surface for action (that is, for elaboration of sap) than those of other deciduous trees, the action of the surface requires to be greater in proportion, to keep the tree in health; and from this, and also from observation of the localities in which Larch thrives, he shows the desirableness of a clear dry atmosphere, with plenty of sunlight, and freedom from fogs and damp which tend to diminish the evaporation from the leaves necessary for the health of the tree; and—to give a single instance—it is noted that at a height near Geneva (not less than that at which fine Larches were to be found) the trees did not thrive near the lake and river, whilst in the dry air of the Alps they prospered.

It appears plain that any cause, great or small, that induces a damp stagnant atmosphere, or want of light round the Larches, will produce ill-health, and in such situations the Larch Bug thrives.

We may do something to diminish the amount of attack by planting on proper soil, and especially avoiding such flat moorish land as is likely to cause stagnation of moisture in the ground; and also by thinning Larch plantations in time, so as to allow as much sunshine as possible on the leaves; and all overtopping by deciduous trees should be carefully avoided.

Hopelessly-attacked trees should be felled and all the twigs burnt, to avoid spread of attack; and in nurseries it might be worth while, besides the dressings given when the Bug is seen to be present, to give one or two thorough drenchings with soft-soap towards the middle or end of August, to deter attack when the eggs for next year's "mother *Chermes*" are being laid.

With regard to such connection as may exist between "Larch Blight" and the diseased cancerous formations known as "Larch Blister," we have no certain knowledge at present.

The causes mainly under consideration as giving rise to this great evil are four :—fungus ; frost-bite ; "Blight," or Chermes-attack ; and deranged circulation of the sap ; and (without venturing to offer a definite opinion) many circumstances appear to point to this last, which may arise from any cause affecting the health of the tree, but especially from weather influences, such as rapid alternations of heat and cold, moisture in the air, and want of sufficiency of sunlight, as at least very possibly the cause of the blister.

Prof. DeCandolle observes that, as an Alpine tree, the Larch is singularly free from disease, and the trunks remarkably healthy ; and that, though sometimes Larches may be seen having a wound of "resinous cancer," it seems to proceed from some accidental cause, such as a blow when the tree was in full sap ; and after noting that he considers the cause of diseases in British Larches must originate from some difference in the physical structure or culture of British or Alpine-grown trees, he observes :—" The want of a sufficiently intense light, owing to the obliquity of the solar rays, and to the opacity of the atmosphere ; and the over damp state of the latter, appear to me permanent causes which, in your climate, must predispose the Larch to a kind of watery plethora."—(See Loudon's 'Arboretum,' vol. iv., p. 2384.)

It has also been observed that a form of blister affects young trees after being transplanted, in which case also the regular circulation of the sap is disturbed.

By examination of diseased specimens (although this is far from affording a complete view of all forms of the disease), the formation of the blister may be traced *backwards,* from the large open diseased wound, to the swelling just bursting, and not yet burst; and then (keeping as a guide the similarity of the diseased spots as shown microscopically) from a patch beneath the bark with no external swelling, to a few small spots connected by a canal, or to a single spot filled with brown disorganised tissue, which appears to me to be the origin of the evil.

In these spots (as far as observable in all the specimens examined by means of a quarter-inch object-glass) there was no trace of mycelium, or of any kind of fungoid presence; after a time, when the blister has become an open wound, it is impossible to say that *Corticium amorphum* (from which "blister" has been conjectured to arise) or other fungoid growths may not be present, just as *Peziza* or other fungi may be found on the bark; but in all the first states of the blister that I have examined, whilst still it had not burst into an open wound, there has not been any such presence.

In these cases—that is, where the "blister" was originating from a mere speck—it differed markedly from the effect of frost-*bite* on twigs of the same age, as the effects of the frost-bite which had then taken place a few weeks before affected the cells in the bark over a surface of several inches, and the condition of the many injured cells in this case, and of the one or few diseased ones in the other, was very different. This observation merely refers to the complete frost-bite, not to effect of weather on health of the tree.

Observations as to the state of the precise spot where the mother *Chermes* has been noticed to be attached by her sucker during oviposition would give much information as to whether any diseased state of tissues was set up by the irritation of suction; when once the disease has taken the form of an open wound, it is very probable that the presence of *many* of the *Chermes* sucking on such young diseased bark as they may find would increase the commenced disease; but the great point is the *origin*.

This appears, as far as specimens show, to be *not a*

growth, but a death; a spot or spots joined by canals filled
with dead discoloured and disorganised tissue, which may
exist for one, or possibly two or three seasons unseen
beneath the bark, until the consequent stoppage of sap
causes a swelled growth, and the diseased mass, composed
of the discoloured cells and passages, and the tumid
swellings, is set on foot; and may be traced forward in
section, increasing year by year from its starting-point.*

* The above remarks on Larch blister are offered with hesitation, as
venturing on a subject where those who have better opportunity than
myself for observations are still in doubt, and therefore I take leave to
mention that they are mainly based on specimens forwarded for examination, or observations on Larch in West Gloucestershire; but not
having the opportunity of studying the subject in the large plantations of
the North, with the thoroughness requisite for a knowledge of the
different developments of the disease, and the different coincident circumstance on which alone conclusions can be based, I merely give these
points as all I have at present to offer.

LIME.

Buff-tip Moth. *Pygæra bucephala*, Stephens.

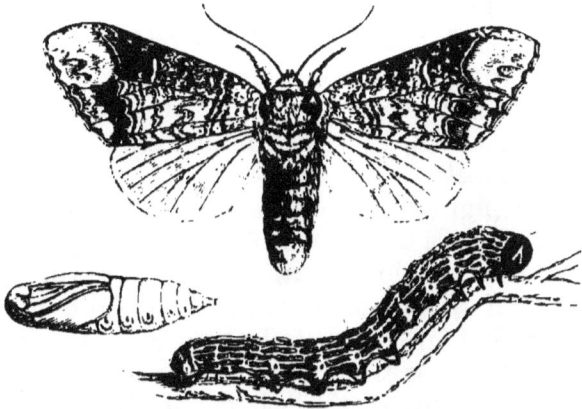

Female moth, caterpillar, and pupa.

The caterpillars of this Moth feed on the leaves of the Lime and also on those of the Elm, Oak, and other trees, sometimes doing thereby serious damage.

The eggs are laid during June or July, in patches of about thirty to sixty, mostly on the upper side of a leaf, and are distinguishable by being convex and white above, smoke-coloured and flat beneath, with a black dot in the middle of the convex part.

The caterpillars hatch in about fourteen days, and at first feed in company on the skin of the upper side and on the pulp of the leaf. After eight days they undergo the first moult (that is, cast the skin for the first time), and separate into parties of eight or ten, which feed at the edge of a leaf, but, when resting, place themselves side by side on its surface. When full fed, which is towards the beginning of autumn, infested trees may be known by the twigs of the higher and outermost branches (or in bad attacks, almost the whole of the tree) being stripped of its leafage.

The caterpillars, when full grown, are about one inch and three-quarters long, and sprinkled with silky hairs; the general colour yellow, with black head, black lines running from the head to the tail, interrupted by a transverse orange band on each ring, and a black horny plate above the tail-segment.

When full fed they come down from the tree, and, without spinning any cocoon, they change at the roots of herbage, amongst fallen leaves, or on or just below the surface of the earth, to a dark brown chrysalis, with two spines at the tail.

The moth (figured above, life-size) comes out in June. The fore wings are of various shades of pearly or purplish grey, with rusty coloured and black markings, and a yellow or buff patch at the tip, whence the moth takes its name of Buff-tip; the hind wings are whitish, with a dusky cloud towards the middle. The head is ochreous, and the body between the wings and abdomen is also ochreous; but variously striped or spotted with more dusky or rust-coloured tints.—('Brit. Moths,' &c.)

PREVENTION AND REMEDIES.—The method of getting rid of the caterpillars that is chiefly recommended is, to shake the infested boughs; it is stated that, on this being done, they fall down "in a perfect shower" (E. N.); and when attack is found to have begun, this plan should be adopted at once.

Any method by which the boughs or infested twigs can be shaken will answer, such as jarring the larger boughs with a pole, or throwing sticks or handfuls of gravel at such parts as may be out of reach; but a better plan would be for a man to go up the tree, and, by means of a strong pole furnished at the end with a worn-down birch-besom, to shake all the infested boughs thoroughly, beginning at the uppermost and working downwards, so as to shake off the caterpillars that may have lodged in falling on the lower branches. The addition of this worn-down stump at the end of the pole makes it a much more effective instrument, for, by using it upright,

the smaller boughs can be lifted up sharply, to come
down with a jerk; or a heavy blow can be given that will
shake all twigs near without any injury to the tree,
whilst in the case of the sharp knock of hard wood
on soft bark given by the pole, much harm is apt to be
done.

Before beginning the operation, a good thick band of
hay or straw, or cloth well tarred, should be put round
the foot of the tree, to prevent the caterpillars getting up
it again; for directly they reach the ground they start
on their return journey towards the trunk, and, unless
they are stopped, will soon be once more at work on the
leaves. All that fall to the ground should be crushed
with the foot, or killed in whatever way may be most
convenient; and where a tree is much infested it would
be worth while to spread large cloths or pieces of
tarpaulin, or anything that might be preferred, beneath
it, upon which they would fall, and from which they
might be collected more easily than from the grass.

The large size of this insect in all its stages and its
habits throws it open to attack. "At the beginning of
June these singular moths may be frequently found
coupled in pairs on the trunks of Lime, Elm, and other
trees, or on the herbage below them; the truncate heads
and closely-convolute wings giving each pair the appearance of a single piece of dead and dried stick."—(E. N.)

By destroying the moths at this stage many future
broods are got rid of. The chrysalids may be collected
by children for a few pence, under or near trees where
the caterpillars have been numerous; and poultry also
are of service, as they will search eagerly for them.

When the caterpillars are about to change, they are so
conspicuous, from their bright colouring, large size, and
habit of straying about everywhere in full daylight, that
many might be captured and killed (as above mentioned)
by children.

As the caterpillars come down the tree to the ground
for their change to chrysalids, it might be worth while
to throw a few spadefuls of gas-lime or of anything they

would not cross, in a circle at about a yard or two *from*
the tree; or a rough band of any material soaked in tar,
or tar and oil, which would keep wet longer, would stop
them from straying off.   It being matter of instinct for
these caterpillars to come down to the ground for the
change to the chrysalis state, probably few, if any,
would return up the trunk, and they might be cleared in
sufficient numbers as to considerably lessen future
attack.

Lime-tree Mite.   *Tetranychus tiliarum*, Müll.
    „   **Red Spider.**   „   *telarius*, Claparede.

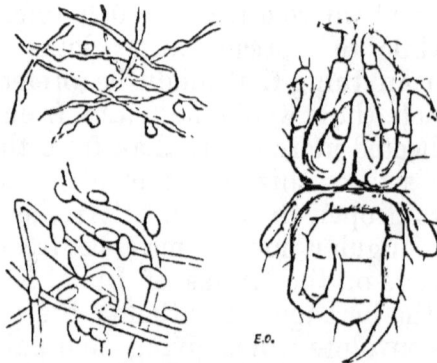

Mite, greatly magnified; web with eggs (magnified), in dried state, and
after being moistened.

Opinions differ as to whether the Red Spider of the
Lime-tree is the common Red Spider, *T. telarius*, or a
distinct species, *T. tiliarum* (so named, from infesting
the Lime), but which is also at times injurious to French
Beans, and some other garden-crops.

These *Acari* or Mites, which are of an orange-colour
and too minute to be clearly discernible by the naked
eye, excepting when heaped together, are of the shape
figured above, enormously magnified, from specimens
infesting Lime-trees at Walthamstow, to which they
were exceedingly injurious in the autumn of 1880.

From their extreme minuteness and transparency, the various parts show very indistinctly when magnified, and the nipping-jaws and sucker were not clearly visible. I have therefore only been able to give the general figure of the mite, with the peculiar long stiff hairs with knobs at the ends (which are believed to help the Red Spiders in spinning) figured on those feet, on which they were distinguishable.

These mites spin their webs over the trunks and branches, and beneath the leaves of the infested trees, giving a kind of glaze or silky lustre to the surface; and on this web they can travel easily. They are to be found heaped like masses of living yellow dust at the foot of the tree, and those on the leaves congregate chiefly on the lower side, sometimes so thickly that none of the green colour of the leaf is visible. Here they draw away the juices with their suckers, and, though they are but small, there are so many of them that the leaves shrink and die from the injury.

PREVENTION AND REMEDIES.—The eggs may be found attached to the webs (see figs.), and it has been found that brushing the tree-stems hard and thoroughly, so as to remove the webs, is serviceable in some degree in clearing attack; and would be still more so if some soft-soap was brushed in at the same time.

With regard to the masses that congregate together at the base of the tree, something might be done by banking round at a few feet distance and a few inches high, and filling the space enclosed with mud made as thin as would be retained by the raised edge. The Red Spider particularly dislikes moisture, and a few experiments would show what chemicals or other additions might be mixed with the mud, to poison as well as drown the pest.

A liberal mixture of fish-oil soft-soap, so as to completely plaster round the foot of the tree and stick all wandering mites fast that touched it, could not fail to do good at a trifling cost; but, excepting by such continuous and thorough drenchings as it is scarcely possible to apply to large trees, it is most difficult to do anything for the infested leaves.

The following recipe, however, might be of service :— Gas-water, three gallons, to which is added one pound of flour of sulphur ; these to be held over the fire whilst being mixed, and soft-soap added in such quantity as to make the mixture adhere. This may be applied to the branches by means of a painter's brush, and where remedies are needed on a large scale it may be diluted to the state in which it is a safe application, and the liquid thrown over the leaves by means of the garden-engine. Probably fifteen parts of water to one of the mixture would be quite safe, but this would require trial.

This attack has been considered to follow peculiar atmospheric conditions, and its severe occurrence at Walthamstow last autumn was after heat and drought, which is generally favourable to Red Spider.—(A. M., F. H., and Ed.)

For life-history of *T. telarius* and general remedies, see Red Spider on "Plum," and on "Hops."

OAK.

Cockchafer. *Melolontha vulgaris*, Stephens.

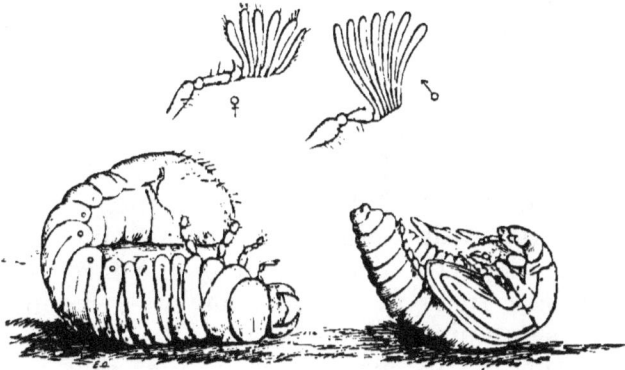

Larva and pupa of Cockchafer. Club of horn of male Beetle (♂) with seven leaves; of female (♀) with six leaves, magnified.

The Cockchafer, known also as the May Bug, is injurious both in the larval and perfect state. As a grub it feeds under ground on the roots of grass, vegetables, and young trees; as a beetle it feeds on the leaves of Oak, Elm, and other trees, sometimes entirely stripping the foliage. The eggs are white or pale yellow, and are laid (early in the summer) about six or eight inches below the surface of the ground, the female burrowing down to deposit them, and laying thirty or more, near together amongst the disturbed earth.

The grubs are thick and fleshy, white or yellowish in colour, with strong jaws, and three pairs of legs; and usually lie on one side, somewhat curved together, as figured above. At the commencement of spring they come up to within a few inches of the surface of the ground, where they feed on roots of growing plants; and at the end of the third summer, when full fed, they again go down into the earth to a depth of two feet or more, and change to pupæ (as figured above) in oval cells.

During the following winter they develop into the perfect Chafers, but do not come up through the ground until the next summer, that is, the fourth year since they were hatched, when they may be found as early as May hanging half-torpid or sluggish beneath the leaves during the day, and coming out on the wing during the evening, when they fly in search of their mates or feed on the foliage of the trees.

The beetle is too well known to require description, but it may be observed it is about an inch in length, densely covered with down on the breast, and more or less throughout; part of the front of the face and the wing-cases are rusty or brown, the latter having five raised lines running along each; the abdomen is prolonged into a tip curved downwards, and marked at the sides with alternate triangular patches of black and snow-white; and the horns are terminated by fans or clubs of seven leaves in the male, six leaves in the female.

This Chafer is sometimes confused with the "July Bug," the "Small" or "Summer Cockchafer" (*Rhisotrogus solstitialis*), but the two may easily be distinguished by the "Summer Cockchafer" being much smaller than the "Common Cockchafer"; also it is hairy, has a blunt tip to the abdomen (not a prolonged tail), and the fan at the end of the horns has only three leaves.

The grub of the Cockchafer is very like that of the Green Rosechafer (*Cetonia aurata*), but is distinguishable by having few or no hairs and by *not* having a rusty spot on each side of the first ring behind the head, which is plainly to be seen on the grub of the Rosechafer.— (J. O. W., in 'Gard. Chron. and Ag. Gazette,' 'Illus. Brit. Ent.,' &c.)

PREVENTION AND REMEDIES.—When the May Bugs or Cockchafers appear in the large quantities sometimes recorded, as when eighty bushels are stated to have been collected on one farm ('Encyc. of Agriculture,' 2nd ed., p. 1166), it is worth while to beat or shake them from the trees, preferring noon-time or early on a bright warm day, when the beetles are clinging beneath the leaves and are dull and sluggish.

They may be shaken down on to large cloths spread beneath the tree, or may be swept together and destroyed, taking care in either case that the Chafers are collected before they have time to recover from the fall and take wing; or, as pigs and poultry devour these beetles greedily, it would save trouble to drive them below the trees and shake the Chafers down to them. Poultry would take some time about the work, but pigs would make a rapid and effectual clearance.

In field or garden cultivation, where many grubs are turned up by the spade or plough, some means should be taken for destroying them. Hand-picking by children is of use, but probably in the fields the pigs would be the better helpers. Their instinct and fondness for the grub makes them hearty and well-qualified searchers.

Wild birds, such as Rooks and Sea Gulls, should on no account be driven off. The Black-headed or Peewit Gull follows the plough in the same manner as the Rooks, and feeds on Cockchafers both in the grub and beetle stages; the Common Gull will go for miles inland to follow the plough in search of insects and grubs; and the Nightjar, by "feeding almost entirely on Cockchafers and moths" during the morning and evening hours, is also of great service.

In pasture lands, where the grass has been seriously injured by Cockchafer grubs feeding on the roots, much service has been done by Rooks turning up the grubs and devouring them : in such cases the birds should be carefully protected from molestation; they pull up little —if anything—more than the infested plant (which would have died), and are in this case almost our only means of clearing off these large grubs, which otherwise, excepting when changing their skins or torpid during severe cold, would continue feeding for three years.

P

In garden cultivation, where the ground is infested, a tame Rook is of much service. The bird will set to work as soon as it is introduced, and keep on steadily at the task, clearing the grubs from spots that could not be reached otherwise without injury to the plants.—(W. S.)

From the circumstance of the Cockchafer grubs feeding amongst roots and giving no signs of their presence till the fading of the attacked plant draws attention to the injury going forwards, it is difficult to find any remedy, excepting by means of the insectivorous birds, which appear to have an instinctive knowledge of the position of the larva below the surface; but where examination of the roots of injured plants shows this grub to be present, it would be worth while to try the effect of good drenchings of some fluid, such as tobacco-water (properly diluted), gas-water, fluid drainings from pigsties, or anything else preferred that would not be injurious to the plant, but would be sufficiently offensive to the grub to drive it from the roots.—('Brit. Birds,' 'Illus. Brit. Birds,' &c.)

Marble-Gall Fly.   *Cynips Kollari*, Hart.

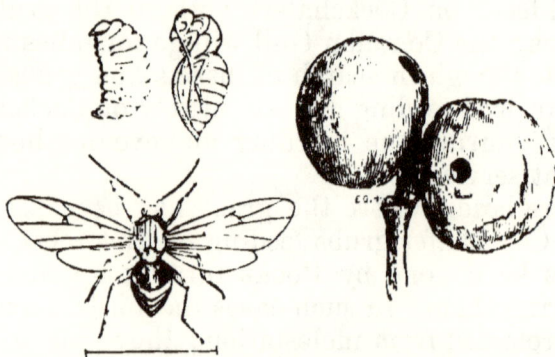

Larva and pupa of *Cynips*; *Cynips Kollari*, magnified.   Marble Galls.

The "Marble Galls," figured above, seldom cause much injury, but they occasionally occur in such large

numbers that they appear to be the most hurtful of the
forty-two or more kinds of Galls which are to be found
on Oaks in Great Britain.

Oak galls infest all parts of the tree; seventeen kinds
occur on leaves, fifteen affect the formation of the buds,
others occur on the bark, root, and catkins, one kind is
to be found in the twigs, and one in the acorn; but with
the exception of the Marble-gall and the "Common
Spangle" gall (*Neuroterus lenticularis*), which sometimes
so completely loads the back of the leaves as to cause
premature withering, it does not appear that any kinds
are materially hurtful.

The females of these Gall-flies (which belong to the
order Hymenoptera) are provided with a peculiar
apparatus for egg-laying, by means of which they are
able to insert one or more eggs with a small quantity of
fluid into the part chosen for attack, and thus set up an
irritation in the living tissues which causes the diseased
growth, resulting in the shape of galls.

In the case of the *Cynips Kollari* the egg is laid in the
young bud when forming in the axil of the leaf, and the
consequence is the globular growth of cellular tissue
which we find (before the Marble-gall is mature), with
the grub lying in the middle: towards autumn this
changes to a pupa, similar to that figured above, and
generally the Gall-fly comes out shortly after, but sometimes not until the following year, or possibly even later.
It is furnished with four transparent wings of the
expanse marked by the line beneath the figure, the body
and abdomen are of a rusty or ochreous brown, and the
base of the abdomen pitchy. As yet females only have
been observed.

This gall of the *C. Kollari* was noticed in such great
numbers in the south-west of England, about the year
1854, as to give rise to an impression that it had then
first appeared in this country; and, from it being
especially observed in Devonshire, the name of "Devonshire gall" was bestowed upon it. Further investigation,
however, showed that its presence had been noted

previously, and that the appearance was only remarkable for its great amount ; now it is widely spread throughout the country, and is to be found as far north as the hills of Bayndee, near Banff, and Redcastle, in Ross-shire.

PREVENTION AND REMEDIES.—This gall is chiefly to be found on low-growing Oak—such as trees stunted by want of shelter from sea-blasts or by other causes ; Oak-bushes in underwood and hedgerows ; the shoots upon the stumps of felled trees ; and also on young trees, and is especially undesirable in Oak nurseries.

It has been recommended to employ children to break off the galls before they are full-grown, and thus destroy the insect within in its maggot state ; but this is but a rough remedy, for the children would probably break off every leaf with the gall attached. It would answer better in nursery management for a man furnished with a common penknife to go through the trees whilst the galls were still young, and with a touch of the knife remove the gall without injuring the leaf. The operation would be an effectual cure, and very rapidly performed. In cases where the galls are formed in clusters of three or four to perhaps eight or ten, it is desirable to cut the shoot off below the cluster. If the galls are cut off whilst they are still soft and young, there is no need to take the trouble to burn them ; they will dry and shrivel, and the maggot within will perish. The encouragement of the Tomtit, or Blue-headed Titmouse, and also of the Black-headed Titmouse, is an excellent means of prevention of increase of these Gall-flies in Oak nurseries. Where galls are numerous, these birds are of great service, by making a rough hole in the gall and picking out the contained maggot.

Other larvæ or maggots, besides that of the *Cynips Kollari*, are often to be found in these galls, sometimes what are known as "inquilines," or fellow-lodgers, which are dispersed in small cells through the substance of the gall ; sometimes parasite larvæ, feeding on the larva of *Cynips Kollari* in the large central cell.

The following list, taken from the 'Gardeners' Chronicle and Agricultural Gazette' for 1868, p. 295, gives the various species and varieties of Oak on which "the gall" was noticed by Mr. J. Barnes as plentiful at Bicton, in Devonshire, but without interfering with plentifulness also of the acorn-crop :—

Quercus	pedunculata.	Quercus	Tauzin.
,,	australis.	,,	Turneri.
,,	dentata.	,,	heterophylla.
,,	pendula.	,,	alba.
,,	pubescens.	,,	montana.
,,	sessiliflora.	,,	pyrenaica.
,,	Louettii.	,,	magna-maculata.
,,	mongolica.	,,	asplenifolia.
,,	xalapensis.	,,	rubra, &c.

The following species or varieties, though growing contiguous to others infested by galls, had never been found to be similarly attacked :—

Quercus	laciniata nova.	Quercus	aquatica.
,,	americana alba.	,,	Phellos.
,,	Hodginsii.	,,	falcata.
,,	palustris.	,,	ambigua.
,,	Chinquapin.	,,	laurifolia.
,,	tinctoria.	,,	coccinea.
,,	Ægilops.	,,	heterophylla minor.
,,	apennina.	,,	heterophylla vulgaris
,,	cerris variegata	,,	macrocarpa, &c.

Common Spangle Gall. *Neuroterus lenticularis*, Ol.

The gall of the *Neuroterus lenticularis* is distinguishable from the four other kinds of Spangle-gall found in Britain by its somewhat larger size, and also by being raised in the centre and hairy. It sometimes occurs in great

quantities on the backs of Oak-leaves, but rarely to an
extent to cause serious damage.

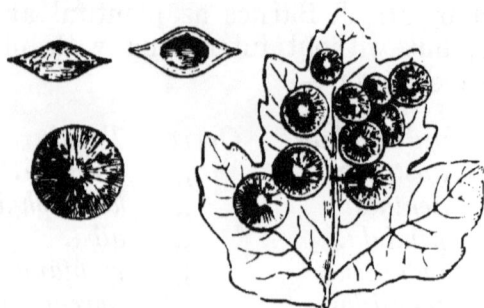

Common Spangle Gall on Oak-leaf, nat. size and magnified; also in
section, magnified.

Oak Leaf-roller Moth.   *Tortrix viridana*, Stephens.

The caterpillars of this moth cause serious injury
from time to time in our Oak woods and forests,
especially in the South of England, by feeding in such
vast numbers on the young leaves as to strip the trees
of their foliage, and thus retard the growth of the first
shoots, and injure or entirely ruin the acorn crop of the
season.

The eggs are laid during the summer or autumn of
the year preceding the attack of caterpillars, either on
or in the leaf-buds, or on the boughs (opinions differ as
to the precise spot); but in the following spring, when
the Oak leaves are appearing, the caterpillars hatch, and
sometimes swarm in myriads over the infested trees on
many acres of ground.

The caterpillars are at first greenish grey, or lead-
colour; when full grown they are dull green with dusky
spots, and about half an inch long.   They have the
power of rolling the tip of the leaf and spinning it
together into a cylinder (as figured), within which, when
full-fed, they turn to chrysalids, but meanwhile, on
alarm, or as matter of choice, they let themselves down

by scores or hundreds, by means of silken threads, for about seven or eight feet, and sway about as the wind may waft them beneath the infested boughs, catching on any passing object, and also a prey to many kinds of birds; but, if nothing else happens, they crawl presently back again, each up its own line to the bough. The chrysalis is brown, and is formed in a silken cocoon on a leaf.

The moth, which appears towards the end of June, is about an inch in the expanse of the fore wings, with the head, body between the wings, and fore wings, of a light

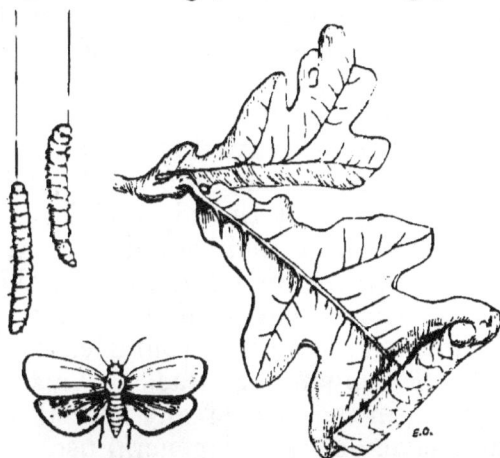

Moth; caterpillars hanging from their threads; and rolled leaf.

green; the hind wings are brownish, and the fringes of the wings, as well as a line on the front edge of the foremost pair, are whitish. — ('Brit. Moths,' Prak. 'Insecten-Kunde,' &c.)

Prevention and Remedies.—From the circumstance of the eggs of this moth being so small as not to be observable on the tree, and also from the caterpillar attack (which sometimes extends over miles of woodland), occurring at irregular intervals without any previous signs to give warning of its approach, it appears impossible, as far as is known at present, to apply any remedy of general service, excepting such as may be found in the encouragement of the wild birds.

In a very severe attack on the Oaks in 1827 it was observed that the Willow Wrens and Whitethroats were useful in clearing the caterpillars; the Chaffinch also; and the House Sparrows "were indefatigable in search of them"; and poultry searched under the trees for such as fell to the ground (Mem. Lit. and Phil. Soc., Manchester, 2nd Series, vol. v.) Rooks, Jackdaws, Thrushes, Starlings, Titmice, Nuthatches, and Woodpeckers are said to be of use in the matter (E. L. T.) Rooks and Jackdaws have been especially observed as flocking to the infested trees, and a second crop of leafage to be soon afterwards established.—(J. H.)

In cases where only a single tree is attacked something *might* be done to save it by sending a man, furnished with a new birch broom, into the branches, where, by starting operations at the top and jarring the boughs, he would cause large numbers of the caterpillars to fall down, and, with the long fine twigs of the broom, could clear away those that floated at the ends of their threads, without hurting the young leaves, and thus much diminish the amount of attack. This plan would be worth trying in the case of any special tree that it was wished to save, but not for general use.

In case of attack occurring on young trees it is probable that drenchings thrown powerfully at the foliage by large garden-engines, would be serviceable. Any of the washes of soft-soap—or soft-soap, sulphur, and gas-water—might be used, for which see Index.

Note. — For Goat Moth (*Cossus ligniperda*), Wood Leopard Moth (*Zeuzera Æsculi*), and Buff-tip Moth (*Pygæra bucephala*), of which the caterpillars attack the Oak, see Index referring to them under the heads of the trees they more especially frequent. Many other insects also feed on the leaves, or in the wood or bark of the Oak, concerning which space does not allow details in the present volume.

PINE.

Pine Beetle. *Hylurgus piniperda*, Curtis.

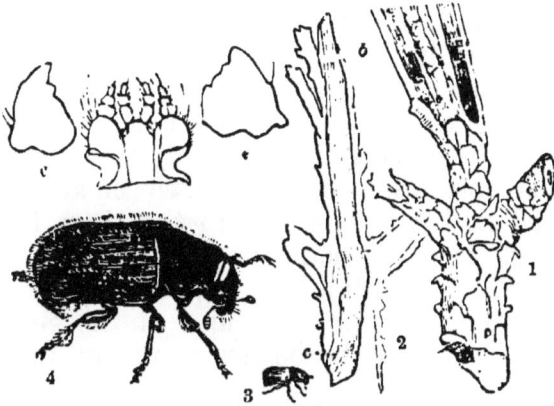

1, 2, Pine shoots pierced by beetles, in section; 3, 4, Pine Beetle, nat. size and magnified; *e e*, jaws; *f g*, chin, with feelers, &c.

These beetles are destructive to Pine plantations (from those newly planted, at all stages of their growth, up to fifty years of age), by boring through the side of the tender shoots into the pith, and eating their way for an inch or two along the centre. This is done in summer, and in the following spring, during high winds, these shoots are blown off. The injury to side shoots by this means is considerable, but in the case of the leading shoot being thus lost the tree often becomes bushy-headed, its growth is retarded, and its ultimate value is reduced.—(W. M'C.) Some amount of injury is also caused by the tunnels which the beetles form in the under side of the bark for egg-laying, but they rarely select healthy trees if sickly ones are at hand, and chiefly frequent fallen wood, felled trunks, or dead or decaying trees and branches for this purpose.

The female appears in April or May, and begins her operations by boring a hole through the bark, beneath which she forms a gallery or tunnel of a little more than her own width ; along each side of this she lays her eggs, from which the larvæ or maggots soon hatch, and each larva eats its way forward beneath the bark, thus forming a series of burrows, gradually getting larger towards the extremities, sometimes running nearly at right angles with the first (or mother beetle's) tunnel, at others bending in various directions (as figured, from near Blairgowrie, R. C.). The burrows are eaten out of the under side of the bark, but often show just a trace of working on the outside of the wood lying against it.

The maggots are about a quarter of an inch long, legless and fleshy, and largest in the rings behind the head, which is of an ochreous colour ; the rest of the maggot is whitish, with a light ochreous tint towards the tail. The maggots turn to pupæ at the end of their tunnels, from which the beetles come out in July and August.

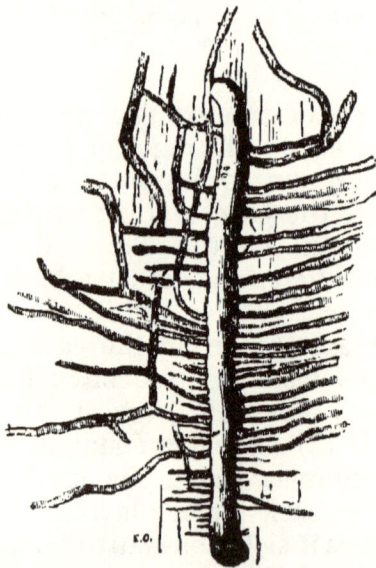

The Pine Beetles are of the size figured above, and of the shape given more clearly in the magnified figure, of a pitchy colour when mature, but paler previously, rough, punctured, with longish hairs, and furnished with strong jaws. The wing-cases are rounded down at the sides, and cover a pair of wings capable of strong flight.

In their first stage—that is, whilst they are still feeding as maggots—they do little harm, this part of their life being rarely passed in healthy trees ; it is after they are developed that the real work of destruction begins. Then they pierce a little round hole through the bark, at the end of their burrow, come out through it, and fly to the neighbouring trees, where they may be found in September in great numbers, boring into the young shoots and injuring them, as above mentioned.

Where the beetles pass the winter is not clearly proved ; it is considered by some observers that they shelter in rubbish and moss on the ground, or in crannies in old bark, and similar places, but it is also stated that they are to be found in standing trees, the resting-place where each beetle has gnawed a burrow into the soft outer wood being marked by a lump of turpentine which has oozed from the wound and hardened outside on the bark. We need more information on this matter ; but however this may be, it is clear the maggots feed and develop under the bark, and this is the point mainly to be looked to for means of prevention.—(' Reports of Inj. Insect Observations,' ' Prak. Insecten-Kunde,' &c.)

PREVENTION AND REMEDIES.—" When young Fir plantations are thinned, all the brush ought to be at once removed, or burned on the ground, as this beetle propagates in the decaying branches in legions. They ascend the standing trees and commit extensive ravages. When Fir thinnings are lotted within the plantations it is a very common practice to dress the bark off to lighten the carriage in transit. The dressing of the bark off should *not* be permitted within the plantation. In a year after I have seen, around these heaps of bark, the ground covered with green shoots blown from the young trees which had been pierced by this beetle."—(W. M'C.) " Pinching off the infested shoots and burning them is the best remedy in the case of small trees. Decaying wood or bark is the favourite breeding-place of the Pine Beetle (and troops of other noxious insects), and these should be systematically collected and burned in Pine woods to prevent the increase of insect pests."—(M. D.) Standing trees that are sickly should be observed, and, if found to be infested, should be felled and removed.

In German forestry it has been advised to bark the trunks and large branches of Pines felled in forest clearings, and thus prevent any mischief from eggs laid by Pine Beetles that have been attracted to the spot. How far this plan may be desirable (or practicable) here does not appear, but the importance of clearing *all dead and decayed timber*, *or Pine-rubbish brush*, which may serve as breeding-places, cannot be too strongly insisted on.—('Reports of Injurious Insect Observation,' 1879, 1880, &c.)

Pine-bud Tortrix Moth.　　{ *Retinia turionana*, Hubn.　　{ *Orthotænia turionana*, Curtis.

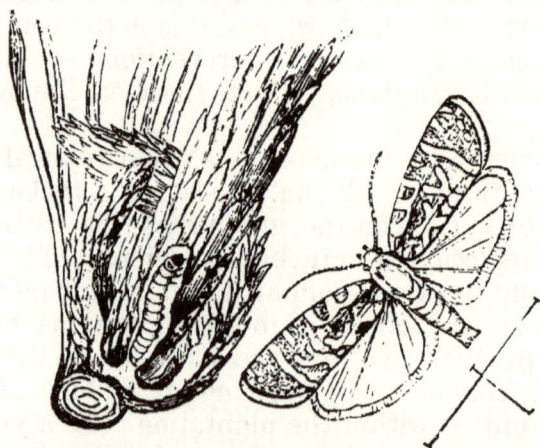

Pine-shoots injured by caterpillars of Tortrix.　Pine-bud Tortrix, *R. turionana*, magnified, with lines showing natural size.

The caterpillars of the Pine-bud Moth are injurious to Scotch Fir, Silver Fir, and various species of Pine, by feeding in the buds, and especially inside the terminal bud of the leading shoot. By this means some of the buds are killed, and the leading shoot is often destroyed and its place taken by a side one, the uniform growth of the branches is interfered with.

The moth appears during July, and lays her eggs on the buds of the young Firs (chiefly selecting those of from five to fifteen years old.—E. L. T.) ; the caterpillars hatch in about twelve days, and feed inside the buds. "At the end of October the caterpillar eats its way from below upwards into the strongest central bud, which is by that time formed for the next year's growth, and there hybernates."—(V. K.)

The caterpillars are usually about half an inch long, reddish or purplish brown, with brown or black head, and with dark bands on the segments ; or with one dark or black band across the segment next to the head. When spring returns they feed again till, at some time between April and June, they change to chestnut-brown chrysalids at the bottom of the chamber they have hollowed out in the bud.

The moth hatches from these in July, and then may be seen resting on the Pine-stems, which it somewhat resembles in colour. The fore wings are from half an inch to a little less than an inch in expanse, and of rusty red colour (sometimes of a darker tint or of a tawny orange), varied with irregular silvery markings (see figure). The hinder wings are whitish grey in the male, darker towards the edge, which has a white fringe ; in the female they are grey throughout.— ('Naturgeschichte Schad. Insecten,' 'Prak. Insecten-Kunde.')

The Pine-bud Moth (above mentioned) and the Pine-shoot Moth (*Retinia buoliana*, mentioned below), which are species of *Tortrix*, resemble each other in so many points of appearance and habits that it is not always possible to ascertain which of the two species is referred to in observations of methods of attack. From the descriptions here quoted it appears that the *R. turionana*.

is the smallest of the two moths, the caterpillars of
which feed chiefly inside the buds, the feeding-season
being both in autumn and in the following spring. The
caterpillar of the *R. buoliana*, which is stated to be the
most injurious of the two kinds, feeds mainly in the
spring of the year on the growing shoots, under a shelter
of threads of its own spinning and hardenedturpentine.

Pine-shoot Tortrix Moth. *Retinia buoliana.*

These Moths are to be found during July about young
Pine-trees of various kinds.

The female lays her eggs between the buds at the ends
of the boughs. The caterpillars which hatch late in the
summer gnaw these so as to cause a flow of turpentine
that gives them a slight coating, and here the caterpillars
hybernate. Their operations are first noticeable in the
following spring when the trees begin their growth, after
which the grubs attack the shoots nearest, or one side
of them, and are to be found sheltered under a kind of
web and the turpentine that flows from the wound.

The caterpillars are at first of a dark brown, which
changes to a lighter colour afterwards; the small head
and a band on the segment next to it are of a shining
black. These are to be found from September to May,
and on ceasing to feed they change (at the same spot) to
chrysalids of a dirty brownish yellow, blunt at the tail,
and furnished on the abdomen with prickle-like
processes pointing backwards. They are to be found
in June in the young shoots, and after lying in this
state for four weeks the moths appear.

These are rather larger than the foregoing species.
The upper wings are reddish yellow, changing to a
darker tint at the tip and marked with light stripes from
the base, and silvery spots and transverse wavy lines;
the hinder wings are blackish grey, with a yellow tint,
and yellow-grey fringes. In the dusk of the evening
they swarm round the tops of the young Pines out of

which they have hatched, but by day they rest and are not readily seen, from their similarity in colour to the withered shoots of which they have been the cause.

This species is common wherever Pine-trees are to be found from the north to the south of Europe.

The infested trees are easily known by the distorted shoots; those that have been injured (and the growth consequently checked) on one side turn downwards, gradually lengthening, till after a while the shoot raises itself upwards at the tip and takes a straight course again; but meanwhile a knee has been formed, and a crippled state given to the branch. The shoots that have been destroyed turn brown and die on the tree, many break off at the bend, and stumpy growths from the number of buds thrown into unnatural development entirely spoil the characteristic appearance of the tree. —('Prak. Insecten-Kunde.')

Prevention and Remedies. — More information is greatly needed; it is noted that a vigorous growth, attention to the trees not being over-crowded, and a suitable soil and situation, are important matters of prevention (E. L. T.); but even the most healthy plantations are not exempt from attack.

Where the state of the buds or shoots shows the caterpillar (or chrysalis) to be present, these should be carefully removed, so as not to injure the remaining shoots, and all these infested pieces should be burnt. This will lessen the amount of future attack, and the earlier it can be done in the season the better, so as to push on a good growth in the healthy shoots that are left by means of the sap that otherwise would have been shared with the infested growths.

From the fact of the moths being sometimes noticeable in large numbers flying in the evening over the infested trees, it is worth consideration whether washings of some kind which would lodge on or amongst the buds where the moths lay their eggs would not be of service to prevent oviposition. These might be applied in nursery ground with an engine, and although tobacco-liquor failed on trial in a plantation in Linlithgowshire much infested by moths, of which the description agrees with the habits of the above species (J. M'L.), it is very likely that some more adhesive wash would be of service (see Index); such as would make a light sticky coating over the buds for the short time the moths were about in large numbers, and which would lodge between them, and so especially protect the spot which the Pine-shoot Tortrix selects for deposit of its eggs.

Pine Sawfly. *Lophyrus pini*, Curtis.

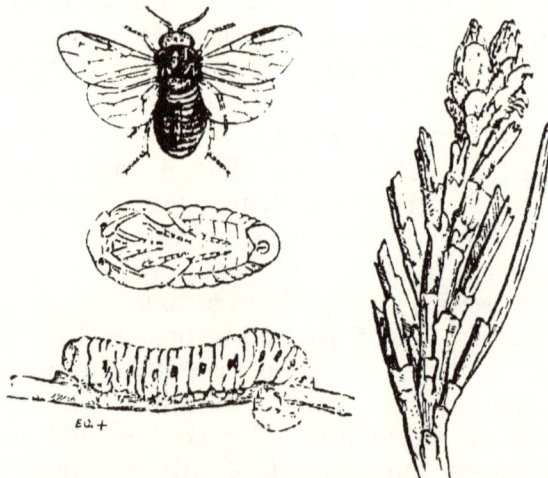

Pine Sawfly, pupa, and larva, magnified.   Pine-leaves injured by Sawfly

The caterpillars of this Sawfly cause great damage to Pines, and especially to young Scotch Fir-woods, by feeding on the leaves. In some cases they scoop away the sides of the leaf, leaving only the midrib; in others, beginning at the tip, they eat the leaves almost down to the sheath. They also feed on the bark of the young shoots, and, as they have voracious appetites and appear in companies, the mischief they do is enormous; and, unless checked by treatment or weather, is continued year after year by successive generations over large areas, sometimes extending to two thousand acres or more of plantation.

The Sawflies appear early in summer, when the female inserts her eggs in the Pine-leaves by cutting a slit along a leaf with her saw-like ovipositor and laying a few eggs in the opening, which she covers with a resinous material scraped from the leaves, repeating the operation until all the eggs have been laid. The caterpillars hatch in about three weeks, and, like others of the genus *Lophyrus*, are 22-footed. They have a pair of claw-like feet on each of the three segments immediately behind the head, the next segment is footless; the succeeding seven segments have each a pair of sucker-feet (or "prolegs"), and the tail is also furnished with a pair, known as the "caudal proleg." The colour varies much with age, health, and weather; at first the grub is green, paler or whitish beneath, with a brownish yellow head, and black sucker-feet; when full grown it has a rusty brown head, dark forehead, and black jaws and eyes; it has an interrupted black line along each side formed of a patch of black dots on each segment; the true feet are black; the sucker-feet are yellow, with a black line at the base; when full grown it is about an inch long.

They feed for eight weeks, and then form cocoons in the moss and leaves or decayed matter beneath the tree upon which they feed, or on the leaves, or in crannies of the bark.

This cocoon is oval, scarcely half an inch long, and small for the size of the caterpillar (which lies doubled on itself within), and is remarkable for the hard compact nature of its exterior.

"The colour of specimens spun under moss is commonly of a dull brown, and of those fastened to the tree either a silky ash-grey, dirty white, or with a yellow tinge; a clean white and a rusty red (the latter commonly with a woolly surface) occur sometimes, but only occasionally."—(Th. H.)

Q

The time taken for development varies; in some
cases the caterpillar remains unchanged for nine months
in the cocoon, sometimes even for a longer time before it
turns to the pupa (figured above, removed from the
cocoon); but the appearance of the perfect Sawflies may
be looked for early in the summer.

The male and female differ from each other both in
colour and size; the male is black, with four transparent
iridescent wings, which are about half an inch in
expanse, and the feather-like rays of the horns are more
developed than in the female. The colour of the female
is whitish, with black head, breast, and horns; a black
patch on the back of the abdomen, and a black patch or
spots between the wings, which are about three-quarters
of an inch in their expanse, and iridescent with purple
and green, varied with yellow, like those of the male.—
('Die Blattwespen,' 'Naturgeschichte der Schadlichen
Insecten,' 'Stephens's Illus. Brit. Ent.,' &c.)

PREVENTION AND REMEDIES.—Clearing away cocoons
from under infested trees during the winter is the best
method of preventing attack in the ensuing season.

A large proportion of the Pine Sawfly caterpillars
which leave the shoots in autumn bury themselves (as
mentioned above) in the dry leaves, moss, or decayed
rubbish beneath the tree, and are stated for the most
part to form their cocoons near the stem of the tree,
where they are sometimes to be found lying together in
masses as large as a man's fist.

"The ground underneath Scots Fir-trees is generally
bare, and covered only with the fallen leaves and
tree débris; so that it is an easy matter to examine
the surface of the ground near the base of the trees,
and, if found infested with cocoons, to scrape it together
and burn it in small heaps, so as to destroy the insects.

"Another plan might be useful; that is, turn over with
a spade the loose surface-soil and tree débris containing
the cocoons, and give it a heavy beat with the back of
the spade, thus smashing and destroying the cocoons.

"However, nothing is so effective as collecting the surface-soil and rubbish into small heaps, and burning or charring it. Even where the surface is covered with rough herbage or heather, this is the best plan, as the rough material will all help the *charring* of the soil, and *burning* of the cocoons."—(M. D.)

With regard to clearing caterpillars off the trees, the following method was found successful on a plantation of about eighty acres near Forres, which was infested by the larva of a Sawfly:—

"When the caterpillars were first noticed, a careful man was provided with a pair of strong gloves, with directions to examine the state of the trees daily, and when he found the caterpillars—which are generally in clusters—to destroy them by infolding the branch on which they were feeding in the gloved hand and pressing it firmly. The caterpillars (which had not appeared in the whole of the plantation, but in great numbers in some parts of it) were thus prevented from doing any great amount of damage."—(D. S.)

Near Dunkeld (where Sawflies had been very injurious for several seasons previous to 1879, on a young plantation of two thousand acres of Scots Fir) an experiment was tried on a small plantation of twenty acres, five miles distant from any other Scots Fir wood, which, up to the date of the observations sent, had proved successful. The plan adopted was to send a number of boys through the plantation, each furnished with a small vessel containing naphtha, and a brush roughly made of feathers, with which the clusters of larvæ were slightly sprinkled or touched, when they immediately fell down, and by this means the plantation was almost cleared.—(J. M'G.)

In the case of a bad attack of Pine-leaf Caterpillars in Roxburghshire, after various means of destroying them had failed,—such as dusting the trees with quick-lime,—the use of hellebore in solution, applied by means of the syringe, was found a deadly application to the caterpillar and an effective cure.—(C. Y. M.)

In the case of larger trees, much good may be done by shaking down the caterpillars and destroying them before they have time to creep away. They fall in great numbers (especially when chilled and slightly torpid in the morning) on the tree being shaken or jarred; and in German Forestry it has been found that one man to shake the tree, accompanied by two women or children with a sheet for the caterpillars to fall on, from which they can be collected and destroyed, can clear fifteen trees of twenty-five years old before nine o'clock in the morning.—(Th. H.) If some fresh Pine-boughs are strewed under the trees before they are shaken, the fallen caterpillars will collect immediately on the sprays, and may be trampled on, or more conveniently shaken on to the cloths to be destroyed, than by simply letting them drop on the cloths from the tree.

It is also desirable, before shaking, to put a band of some nature that the caterpillars will not cross on the ground at the foot of the tree, to keep all that may have escaped from making good their return up the trunk.

Quick-lime would answer this purpose, or gas-lime; or a hay-band (or pieces of any old rags twisted together into a rope) well tarred or soaked in a mixture of tar and oil that would keep wet and sticky for some time, would be a sure preventive of traffic of the caterpillars across it.

When the caterpillars have consumed the leafage on one tree, they migrate to another, and where tracts of forest are affected it has been advised to dig ditches not less than two feet deep and two feet broad, with the sides as perpendicular as possible. Looking at the clinging powers of the caterpillars, it does *not* seem likely that this plan would do more than delay progress, and also afford a clear space where the caterpillars, when they occur in the myriads described by Hartig and Kollar in the German Pine-forests, might be duly dealt with by regular watchers; but, generally speaking, a broad band of something which they would not cross laid on the ground, appears a more practicable remedy.

Sand, or ashes, or dry earth, well sprinkled with paraffin and water, would probably check the onward progress more effectually at less cost; or a band of fresh gas-lime would be effective. Quick-lime would be of little use in this case, as something is needed of which the effects would last for at least a few days.

When infested and uninfested trees are mixed together, it may be worth while to isolate such as have not been attacked. Save where the boughs touch, the caterpillars can only reach them by crawling up the trunk, and a large number might be protected at a small expense by placing rings of any deterrent the forester might choose at the lowest part of the trunk, or on the ground round it. A band about a foot wide of fish-oil soft-soap, mixed to a thick consistency and laid on with a large brush, would cost little beyond the wages of the operator, and probably be a preventive.

Something may be done by picking cocoons off the leaves, or clearing them from crevices on the bark; or by removing the leaves that have eggs laid in them; but these operations are not practicable on a large scale, and (looking at the susceptibility of the caterpillars to injury from wet and cold, in their young state and when they are changing their skins) something might be done in attack on trees in nursery ground, and on limited areas, by syringing. The kind of engine used in Hop-grounds with a double hose would be applicable to the work, and, where water was at hand, the rapid clearing it would be almost certain to effect would be worth a trial.

Ungenial weather acts powerfully on this insect. In the autumn of 1880 it was noted that the first frosts, coming suddenly, destroyed many of the caterpillars that still remained on the trees at Earlston, in the south of Scotland (W. W. R.); and the absence of the Sawflies from the Athol forests, and also their almost total disappearance from the young plantations in the neighbourhood of Beauly in 1880, after having ravaged the young Fir-woods for five or six years until acres were completely stripped of their leaves, is attributed to the low temperature of the preceding summer.—(D. D.)

We have some help in keeping down this pest both from birds and the smaller Mammalia. Woodpeckers of various kinds, Jays, Cuckoos, Titmice, Hedgesparrows, and Swallows, are of service in destroying the perfect Sawflies and also the cocoons, with the contained caterpillar; but they shun continuous attack on the caterpillars on the trees, and diet on them appears to be prejudicial to the nestlings. Amongst four-footed enemies the Field Mouse and the Short-tailed Field Mouse (Field Vole), both of which when driven by hunger are carnivorous, are stated to destroy cocoons lying beneath the moss, together with their contents. Squirrels are not less destructive, as many as a hundred of these spun-up caterpillars having been taken from the stomach of a single specimen; but they will not eat the caterpillars whilst feeding on the leaves.—(Th. H.)

The surest methods of prevention of these pests, however, appear to be in taking advantage of their habits of forming their cocoons in large numbers beneath their food-trees, and of falling from the branches on a sharp shake being given to the tree.

Giant Sirex.   *Sirex gigas*, Linn.

The maggots of this beautiful Fly are injurious to Fir-timber by boring galleries in the solid wood. The method and amount of injury are variously estimated, but by comparison of the observations of many writers it appears that the female *Sirex* lays her eggs in various kinds of Pine—Scotch Fir, Silver Fir, and Spruce—which, though not decayed, are not in full health; such, for instance, as trees past their prime, or that have been uprooted, or broken by wind or accidents, or are sickly from any other cause.

" Zinke and Bechstein agree that an insignificant local injury to the tree affords a point for attack; that the females lay their eggs on such damaged spots, from which the brood spreads, and thus in a few years an otherwise healthy trunk is destroyed."—(Th. H.)

The eggs are also stated to be deposited in felled Fir-trunks left lying in the woods.

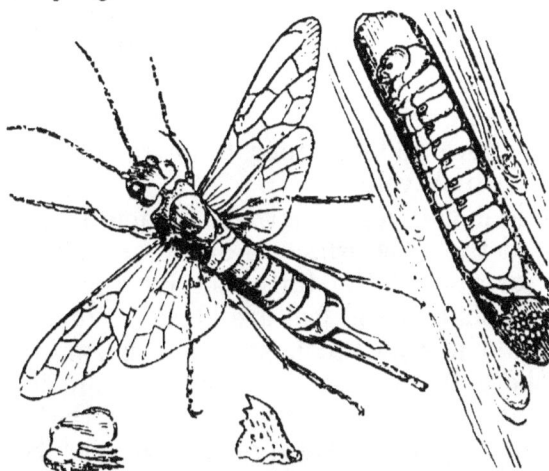

Female *Sirex ;* larva. Jaws of larva and fly, magnified.

The female (as figured above) is furnished with a long ovipositor, by means of which she bores a hole through the bark of the stem of the tree for the deposit of her eggs.

The maggots from these are whitish, soft, and cylindrical, with a scaly head armed with strong jaws ; a blunt point on the tail-segment, and they have three pairs of very minute feet. These larvæ feed in the solid timber, and are full grown in about seven weeks ; and then or later (for how long the larval and pupal state last seems uncertain) they change to chrysalids in the tree. The pupa resembles the perfect insect lying still, soft, and white, with the limbs laid along the breast and body.

The further change to the complete insect may occur in a month, but if the maggot has not turned to the chrysalis till autumn the fly will not appear till the following summer, or even a much later period.

The female *Sirex* is usually an inch and a half long, cylindrical, and with the head and the rest of the body of the same diameter. The colour is black, banded with yellow on the first two and last three rings of the abdomen, and there is also a yellow spot on each side of the head. The abdomen has a short blunt point at the tip, and underneath it is furnished with the ovipositor, which is long and black, and lies in the yellow sheath shown in the figure. The thighs are black, the shanks and feet yellow, and the four large membranous wings are of a brownish yellow.

The male is smaller, with the abdomen flat and yellow, excepting at the base, which is black, as well as the last segment (or end of the tail) and its appendage. The hind pair of shanks and feet are black or dusky; pale, or with yellow rings at the base. The horns are yellow; those of the male are nearly as long as the body, those of the female are rather more than half that length.— ('Die Blattwespen,' 'Naturgeschichte der Schad. Insecten.')

PREVENTION AND REMEDIES. — The best methods of prevention are to clear away trees that are in a condition to attract attack—such as trees that have been injured by accident or ill-treatment, or that are weakened by disease or attacks of other insects, and also those that have been blown over or that have been felled, as the *Sirex* lays its eggs in felled as well as in standing timber.

Any trees which are found to be infested (either by the *Sirex* being seen escaping, or by the large holes in the trunk showing the escape to have taken place) should be felled and disposed of according to their condition, so as to stop further spread of the insect from them.

If they can be taken to the saw-pit and converted to any rough use it is best, for thus the infested parts may be cut off and burnt, and the sound timber preserved; but if this cannot be managed, something should be done, both with trees in this state and felled trunks lying in the woods, to prevent the insects escaping.

If nothing else occurred the tree might be split for firewood to be used at once, or it would be worth while to heap up any rubbish near over the trunk and char the outside.

Sometimes the insects appear suddenly in great numbers. I have seen twelve to twenty specimens captured in a few hours, as they came out of one Larch-trunk lying by a Fir-plantation in West Gloucestershire (Ed.); and in such a case a child with a net could easily catch and kill them. Generally, however, they appear singly or a few at a time, often over a period of several years from one trunk.

Steel-blue Sirex. *Sirex juvencus*, Linn.

This *Sirex* is said to be "decidedly the most common, at least in this country, of the genus" (J. F. S.); but, as its habits resemble those of the Giant *Sirex*, that species has been selected for illustration on account of its great size and beauty.

The *S. juvencus* is often much smaller than *S. gigas*, and generally of a blue-black, varied with bright or rusty red. The colouring, however, is very variable. Sometimes it is blue-black, with rusty red thighs, and reddish shanks and feet; black horns, and somewhat transparent brownish wings, with rusty veins and spot on the fore edge. Sometimes, however, there is a larger amount of bright red marking; the abdomen is bright red, save the two rings at the base, or the six lowest joints of the horns are red instead of black, and many other small differences of colouring also occur in different specimens. —('Illus. Brit. Entom.')

Pine Weevil. *Hylobius abietis*, Stephens.

The Pine Weevil is injurious to Scotch Fir, Spruce, Larch, and some others of the Coniferæ, by feeding on the tender bark of the young shoots. It mainly attacks young trees, especially plantations formed on ground from which a crop of old Fir has recently been removed, and eats away the bark of the stems, sometimes completely stripping them upwards. It also eats the bark of the shoots, and destroys the bud; and, in the Larch, it gnaws at the base of the leaves so as to render the shoots bare.

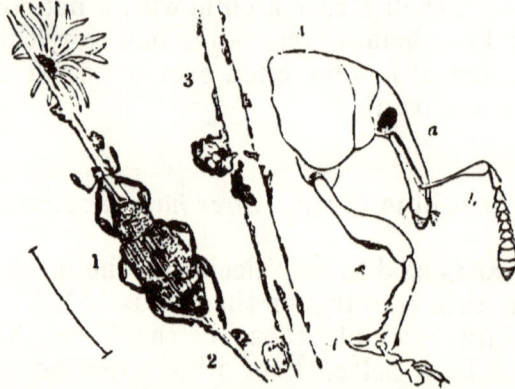

1, Pine Weevil, magnified; line showing nat. length (snout included); 2, 3, Larch twigs injured by Weevils; 4, head, with snout and horn and fore leg, magnified.

The beetles appear early in the summer, sometimes in May, but chiefly in June and July. In unfavourable weather they remain under shelter of the leafage, but when it is warm and sunny they are more active, and pairing then takes place.

The females deposit their eggs, which are transparent and whitish, in rifts of the bark, in logs, root-stocks, stumps of felled trees, and on exposed parts of roots.

The maggots hatch in two or three weeks, and may be found from June onwards throughout the winter. When full grown it is about half an inch long, fleshy and white, with a brown head, which, as well as some portions of the maggot, is beset with bristles. It is either legless, or with mere indications of legs on the three segments behind the head, and in general shape resembles other weevil maggots (excepting that the three segments above mentioned are so much enlarged as to give a swollen appearance to this part of the maggot, which is also much wrinkled transversely), for figs. of which see references to "Weevil" in Index.

The maggots form more or less winding galleries in the soft wood beneath the bark, which gradually increase in size with the growth of the maggot, and, following the course of the root, go down to some depth below the surface. These galleries are gradually filled with the results of the wood-gnawings ("worm-meal") left by the maggot, and at the extremity of the boring there is a cocoon-like accumulation of chips forming a nest for the pupa.

The pupæ resemble the beetles in shape, but with the legs and partially developed wings and wing-cases, and also the long snout or proboscis folded under them; the rings of the abdomen are slightly saw-like at the sides. These pupæ are to be found in spring in their cocoon-like nests, and in this state they lie quiet for about four weeks, when the young beetles develop and come out, whilst some of the beetles of the previous year that have hybernated are again to be found. These latter have passed the winter in moss, or fallen leafage and twigs, or even in holes in the earth, or roots under the trees, or similar sheltering places, and may be known from the freshly-developed beetles by their more faded and worn appearance.

The beetles are about half an inch in length, black, with some patches of yellowish hairs on the head and on the body behind the head, which is also thickly and deeply pitted. The wing-cases are rounded at the sides, and bluntly pointed at the tail (so as to be somewhat boat-shaped); they have alternate lines of punctures and tubercles, and are variegated with spots and bands of yellowish hairs. The legs are black. The beetle is especially distinguished by its long snout or proboscis, with elbowed horns, of which the long lowest joint fits into a hollow formed for its reception in the side of the proboscis (see fig. magnified).—'Prak. Insecten-Kunde,' 'Illus. Brit. Ent.,' &c.

PREVENTION AND REMEDIES.—One of the chief points
to be observed in the habits of this beetle is that it
frequents forest clearings, that is, spots where Fir trees,
few or many, have recently been felled.   Here it
harbours under pieces of bark, broken wood, &c., and
lays its eggs on the logs, stumps, and exposed roots.

It is therefore desirable that all such points of
attaction should be got rid of out of the wood.   There is
no occasion that anything should be wasted, for the
fragments that are only useful as fire-wood may be burnt
long before any eggs laid in them can develop through
all their transformations, but they should *not* remain in
the woods.   Fragments of roots left in the ground
should be thoroughly covered with at least six inches of
earth, and *no* logs (which are an especial haunt of the
beetle for oviposition), should be left about, unless some
of them, or of the fragments of bark, are used for
traps.   This has been found practically useful in German
forestry, and if these traps were regularly examined,
and the contents destroyed, they would probably be
an excellent means of getting rid of many of the
weevils, which will frequent a clearing so long as it
is in progress, and the air pervaded with the scent of
turpentine.

Bark-traps are made by laying pieces of bark with the
inner side downwards on the ground.   Pine bark is said
to answer better than Spruce, as it remains fresh longer.
The pieces should be well weighted down with stones,
and examined early or late in the day.   The number
of beetles caught is greatest in dull weather or during
soft rain.

Log-traps make a good decoy to attract the female
beetle to lay her eggs ; in this plan thick pieces of logs
with bark on them are partly buried in the ground.   The
beetles resort to them, feed on the sap, and lay in them,
and numbers of beetles may thus be caught and the
brood from the eggs got rid of afterwards by burning the
logs when convenient.

Brush-traps are made of Pine or Spruce twigs, tied

together in bundles, about the size of a birch-broom ; these are scattered about infested spots, and attract many weevils, which may be easily shaken out of them and destroyed.

It cannot be too strongly insisted on that all such places for propagation as dying trees, waste timber, broken bark, stocks, roots, &c., that cannot be removed, buried, or utilised as traps, should be carefully gathered together and *burnt.*

The neighbourhood of forest saw-pits should be similarly attended to, and any logs of felled timber, whether in or out of the woods, should be observed relatively to their serving for breeding-places. In the summer the beetles may be found pairing on planks, cleft timber, &c., far from woods.

It is, however, not worth while (unless special circumstances should point out otherwise) to meddle with rubbish under Pine trees, such as moss, dry Pine-needles, small twigs, and such like ; the Weevil does not feed on these, and does not lay in them (for they would not afford food to the maggot) ; it only shelters itself in them for the winter, and the small number that would be destroyed would scarcely repay the labour.

The second and most important point in the habits of the Weevil is that of attacking young plantations, and especially those that have been formed on recently cleared Fir forest lands. It has been found that on land cleared of Scots Pine and planted within four years with Coniferæ, that there would be few remaining uninjured in two years after.

The following method of treatment has been thoroughly successful :—

After the Pine crop is cut and cleared the ground is properly enclosed, so as to exclude stock of all kinds, and, if required, it is drained. The ground is permitted to rest the first summer for the purpose of getting up all herbage as strongly as possible, and in dry spring weather the whole is burned, so as to destroy the eggs and food of the beetle, and as far as may be stamp it out.

After this the ground should be planted with strong two years' transplanted plants.   After each young tree is planted a layer of earth is laid round about two inches in thickness and eighteen inches in diameter ; this layer should be beaten smooth with the back of the spade to prevent the beetle lodging under any part of the rough surface.   This treatment was found to answer well, for as soon as the beetle in search of food comes in contact with the bare earth it immediately steers its course in another direction, and leaves the plant untouched. The beetle is most destructive in dry warm seasons.— (W. M'C.)

Following up the same principles with different details, it has been found a good plan, where planting has to be done on a large scale, and the beetle is present to any extent, to take out as many of the old roots as possible, burn all the rubbish that is lying about, and graze the ground with cattle for three or four years before replanting.—(W. W. R.)

In the first method of treatment, allowing the growth of grass, and firing it (as is observed), " stamps out " the attack ; in the second, the turning the land into grazing ground for " three or four years," gives time for all eggs or maggots to have passed through their transformations before replanting, and the presence of the cattle and their droppings on the ground are an excellent preservative from weevils being attracted to the spot.*

The following remark is also valuable :—" I strongly recommend surface burning when successive planting is contemplated.   This is the safest method of destroying

---

* In a young Fir wood of about eight acres in West Gloucestershire that it was wished to keep in trim order in connection with pleasure-grounds, it was the custom yearly to "skirm" (i.e., rough-mow) the ground with briar or common scythes, and fire the rubbish in a large number of small heaps piled amongst the trees, and, where convenient, on the stump of a felled Fir.   When proper attention was paid the fire did not run, and this surface treatment, together with the general slight smoke from the smouldering heaps throughout the plantation, was accompanied by (if it did not cause) an absence of Pine Weevils which some years before had appeared in great numbers at the locality.—(Ed).

the Fir Weevil in its various stages. If for special reasons the surface cannot be burned, it is well to delay planting for a few years until the weevils are exterminated. —(D. S. S.)

With regard to remedies that are applicable when Weevil-attack is present in young plantations of limited extent, hand-picking is a sure but a tedious and expensive cure. The beetles may be gathered into wide-mouthed bottles (J. M'L.), or they might be shaken down on to tarred boards (see " *Otiorhynchus.*")

Quick-lime has been found to answer well, when thrown on the ground round the trees, and from the observations of the beetle attack, commencing at the ground level and gradually stripping the trees upward of their bark, it would appear that any deterrent thrown round the stem would be useful. Ashes or sand sprinkled with diluted paraffin, or gas-lime scattered round the trees, would probably do much good, but, though the beetles occur to a certain extent on the ground, their low and short flights exactly suit them for attack to the young trees, to which they are mainly hurtful; and mere isolation at the roots can only be looked upon as a partial remedy.

The plan of painting over the stem of the young trees with a mixture of paraffin and red-lead has proved fairly successful (up to the date of the note of observation), but was not considered by the experimenter to have then been tried long enough to be certain of its success or its effects on the tree.—(W. W. R.) It has been found that young trees dressed with a mixture known as Messrs. Davidson's composition, used to keep off rabbits, have been free from Weevil-attack.

From these notes it is plain that direct applications to the stems of the young trees are serviceable, and it would be well worth trying whether applications of soft-soap and sulphur, or of gas-water, soft-soap, and sulphur, well laid on with a brush, would not answer (for recipes see references in Index). Probably smearing the stems with a mixture of cow-dung and lime would do good, or in this case, as the application is not to a food-crop, there would be no objection to using some of the regular insect poisons. Paris Green (Scheele's Green) is retailed at sixpence per pound, and, as an arsenite of copper is a virulent poison; under proper superintendence it would be worth a trial, either laid on dilute with a thick lime-wash or mixed with fine ashes in the proportion of at least twenty of ashes to one of the arsenic green.

One important point yet remains; it appears that the young plants are most attacked after transplanting, and probably in this case, as in many others, the temporarily altered state of the sap attracts the insect feeder, and (as in other cases) all possible care to avoid what would prolong this state, or cause a sickly growth, should be afforded; and when a large extent of ground is covered with Pine plantations, a *very strict* supervision of the method in which the labourers put in the young trees could not fail to do much good.

Spruce-Gall Aphis. *Chermes (pini) abietis,* Linn.

The Spruce-Gall Aphis, known also as the Spruce *Adelges,* or Spruce *Chermes,* causes the small bright green, or green and rosy, galls, shaped like miniature Pine Apples (or somewhat like Scotch Fir cones furnished on each of the divisions with a short leaf), which may be found not unfrequently in the early summer forming at the ends of the shoots of the Spruce Fir.

The mother *Chermes,* from which the brood of the year originates, is very similar in shape to that of the Larch, but rather smaller, oval, wingless, and woolly, of various shades of green or purple, with dark legs, and may be found in spring with her sucker inserted in the base of a Spruce bud, thus causing the irritation which starts the diseased growth known as the "Pine Apple Gall" or "Pseudo Cone." Sometimes the shape is perfect, but often only one side of the shoot is swollen, and the other is merely stunted.

The first growth of the gall and the first egg-laying of the *Chermes* begin in May, or sometimes in the later part of April, and deposit of eggs goes on slowly, the *Chermes* never stirring from the spot during the time, till, having laid a mass amounting to about two hundred, of various tints of yellowish green or grey covered with wool from her own body, she dies.

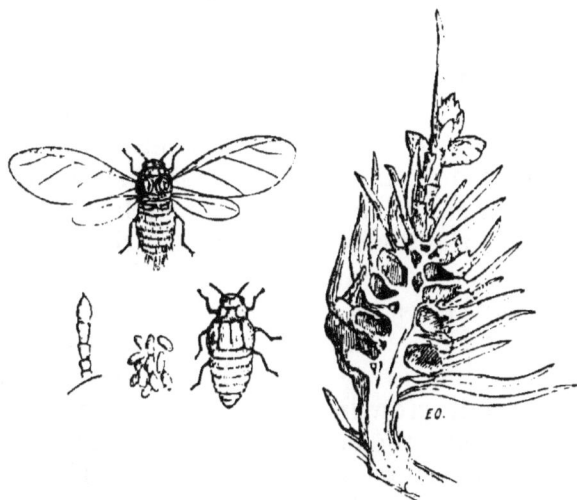

Winged female, pupa, eggs, and horn, all magnified. Section of *Chermes* gall in dried state after departure of *Chermes*.

The larvæ, which hatch shortly from these eggs, are mere specks in size; when magnified they are seen to have six legs, and a head with horns, and to be in shape much like the pupa figured above (at first without signs of the future wings). The colour is greenish, or of a red tint. Meanwhile the growth of the gall—the "pseudo cone" as it is called—has been continuing, and the young *Chermes* larvæ spread themselves, soon after hatching, on its surface, drive their suckers into the soft substance of which it is formed, *and, according to various observers*, become buried in it from the continued enlargement of the base of the unnaturally swollen leaves, of which the gall is chiefly composed, gradually *overlapping*

R

them. This point is one of much interest. With regard to the larvæ that were hatched *outside* presently becoming *tenants of the inside* of the gall, there is no doubt, but, according to my own observations after long and careful watching of the growing specimens, I believe that at a certain stage of the growth of the gall a minute slit opens along the upper part of the sutures that mark the divisions of the swollen leaves of the gall from each other, and through these openings the larvæ creep into the chambers within.

On this point enquirers may satisfy themselves by watching the *Chermes* galls at hatching time, with the help of a strong magnifying glass, especially on the afternoon of a sunny day, and noting (should the process coincide with the above observations) the larvæ spread themselves along the lines which divide the galls into diamond-shaped scales, apparently piercing into them with their suckers, and then disappearing into the chambers of the gall. When this "pseudo cone" has reached its full growth, which may be in four to six weeks, it hardens, the cells split open, and the contained young *Chermes* come out in numbers. The pupæ are powdery, lead-coloured, and margined and greenish at the sides from the indications of the tint of the coming wings. When fully developed the skin cracks, and the perfectly-winged insect appears—figured magnified above (the natural length, at rest with the wings folded, is about the eighth of an inch). The colour is of a yellowish green, with whitish green wings, transparent green legs, and five-jointed horns, also of transparent green; sometimes the colour of the insect is reddish.

The winged females disperse themselves and begin to lay, and soon may be found dead by their little heaps of about twenty eggs. "The larvæ which hatch from this second deposit of eggs" . . . "are, in the next spring, the mother *Chermes* of the attack of the current year."— (E. L. T., Prak. 'Insecten-Kunde,' 'Forst-Zoologie,' 'Gardener's Chron.,' Ed., &c.)

PREVENTION AND REMEDIES.—When Spruce trees in young woods are much infested they should be felled, and, if cut down in summer whilst the galls are green, all gall-laden shoots should be cut off and burnt. In winter this precaution is not needed, as the old galls are empty, and, if the mother *Chermes* should lay on the felled shoots, the buds would not develop enough to nurse up the young brood. This clearing of mature much-infested trees is very important,—cure is hopeless when they are in this condition,—and whilst each year they become more unhealthy under the attack, they form centres to spread the *Chermes* all around.

Where young trees only a few feet high are attacked, it is desirable to go over them and remove the gall *carefully*, so as not to bruise or tear the other shoots, and it is well to do this as soon as the galls begin to show. The sap that would have gone to the distorted growth is thus preserved for the healthy shoots, and hatching of the *Chermes* out of the galls is prevented. When growth is more advanced their removal is best performed by a man furnished with an apron with a large pocket; into this each gall should be put as it is cut, and the collection should be most carefully destroyed. An apron is more convenient than a basket, which requires the use of the second hand; but if the galls, save in their earliest stages, are merely thrown to the ground, the *Chermes* will develop within, and probably be in no way checked by the operation.

How far soil and situation affect the amount of attack does not appear to have been fully noted, but probably they have the same influence as in other cases of Aphis attack. The worst instances of gall presence that I have seen were on trees of about thirty years old, which were somewhat overcrowded and in a damp locality, on a cold, stiff clay; and also, *after removal*, on some fine young trees about three or four feet high, which had been planted in a space in a Fir wood so sheltered by the neighbouring trees and hedges, and also by long rough grass and weeds, that there was no free play of air.

Where there are only a small number of young trees
to be attended to, drenchings with any of the Aphis
washes in July, or when the *Chermes* were seen to be
hatching, would be useful in clearing many from the
trees.

PART III.

FRUIT CROPS

AND

THE INSECTS THAT INJURE THEM.

FRUIT CROPS

AND

THE INSECTS THAT INJURE THEM.

———◆———

APPLE.

American Blight.
Woolly Aphis. } *Schizoneura lanigera*, Hausm.

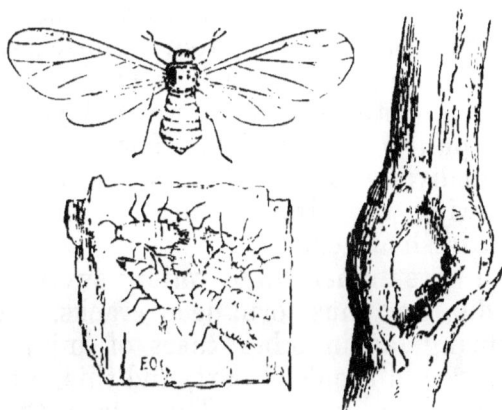

Winged Woolly Aphis, magnified; larvæ much magnified. Apple twig, with the same larvæ nat. size at the lower part of the infested spot.

THE attack of the Apple-bark Plant-louse, or Woolly Aphis, commonly known as American Blight, may be easily detected by the woolly or cottony growth on the

insects, giving the appearance of a white film growing at
the bottom of the crevices where a few of them are
lurking. Where there are many the spot appears as if
a knot of cotton-wool was sticking to the bough, or even
hanging down in pieces several inches in length, ready
to be wafted by the first gust of wind, with all the insects
in it, to a neighbouring bough.

The "Blight" is chiefly to be found in neglected
Apple orchards. Its head-quarters are in crevices in the
bark, or in hollows where young bark is pressing forward over the surface where a bough has been cut off, or
broken by accident so as to leave a shelter of the old
dead bark outside; it may, however, be found on almost
every part of the tree into which the Aphis can pierce
with its sucker; and the harm caused by the attack is
not only from the quantity of sap drawn away from the
bark or young shoots, but also from the diseased growth
which is thus set up. The bark is at first not much
affected by the punctures, but the woody layers beneath
become soft, pulpy, and swollen. The cells and fibres
divide and subdivide, and the bark splits open over the
swelling, showing the tissue beneath, which is thus
exposed for a fresh attack.

At the end of summer these watery swollen growths
dry up and die, and thus form deep cracks. With the
return of spring (as in other cases of injury) a new
growth forms round the dead part, and this soft tissue is
ready for the young Aphides. Thus, from the swollen
diseased growth caused partly by the Aphides, partly by
the natural attempts of the tree to repair damage, a
constantly increasing diseased mass arises which shelters
the insects in its crannies, and finds food for them in its
young hypertrophied formations.

The above note of the progress of diseased growth
is from the paper by M. Prillieux in 'Comptes Rendus'
for April, 1875.

The "American Blight" Aphis is stated to have been
imported from America in 1787, but whether this is
a fact appears somewhat uncertain. It may be known

at a glance from the common Apple Aphis (scientifically *Aphis mali*), which is injurious to the leaves, by the white wool with which it is more or less covered, and from which it takes its name of "Woolly Aphis," and examination of the wings through a magnifying-glass will show that they are differently veined. A strong vein runs down the fore wing near the front edge, and from this three veins turn off towards the hinder edge. The *third* of these veins from the body has only one fork in the American Blight or Woolly Aphis. By this the *Schizoneurinæ*, to which division it belongs, are distinguished from the *Aphidinæ*, which have *two* forks to this vein (as in Hop Aphis); from the *Pemphiginæ*, which have this third vein *without* a fork (as in Lettuce Aphis); and from *Chermisinæ*, in which this third vein is absent (as in Larch Aphis and Spruce Aphis). This difference in the veins of the fore wings is one clear distinction between the above-mentioned four tribes, of which the great family of *Aphididæ* (which includes all the various kinds commonly known as *Aphides*) are composed.

The Woolly Aphides are without honey-tubes, and underneath the wool are mostly of a yellowish, reddish, or reddish plum-colour. The winged specimens are described as pitchy between the wings, and green, or with the abdomen of a chocolate-brown. The wingless females may be found packed closely together in the cottony masses, with the pale reddish young (similar to those figured above, much magnified, from specimens taken near Isleworth during the winter), moving about amongst them. Winged specimens may be found in July and August.—('Prak. Insecten-Kunde,' 'Mon. of Brit. Aphides,' &c.)

PREVENTION AND REMEDIES. — The great harbouring points of this Aphis, and the nooks from which the broods come forth in spring to infest the trees, are crevices, especially such as are formed of young bark sheltered under old dead masses. It is therefore very important to keep up a clean, healthy, well-trimmed state of the branches, such as will not allow of lurking-places, or, if they do exist, will allow of these points of attack being carefully watched. Boughs must be removed in pruning sometimes, and where the Woolly Aphis exists it is certain to try to effect a lodgment under the ring of young bark that comes rolling forward over the stump, but an eye to this matter and a few strong soap-suds brushed on the first bit of wool seen will keep all right; whilst on trees with the boughs maimed by beating the crop off, bad pruning, pieces torn off by the wind, &c., the Aphis gets such a hold in the rough bark as can hardly be got over.

For the same reason the bark should be kept clear of lichens and moss, which form excellent lurking-places for the Aphides. It is also very desirable to keep the trees from touching each other, to keep the ground below them in decent order, at least not totally overgrown with weeds, and to have the soil properly drained. By this means light will be let in, with a free and healthy circulation of air, and the insect-feeding birds will have a chance at getting at the " Blight," whereas in many of the orchards in the West of England, where the principle is held "trees should touch," the tops look well enough to the passer-by, but there is a different story beneath. Here a pale green light struggles through the thick canopy of leaves, nettles and rubbish are often knee-deep, and the limbs of the Apple trees are disfigured with the blight and tumours of many years' growth, and hung with tassels of the wool of the Aphis which show its presence in destroying hosts. A clean, healthy bark, with a proper allowance of air, light, and drainage, is the best of all means of prevention.

With regard to remedies :—The colonies of insects remain in one place, and soon die if their food is cut off or their breathing-pores choked; so that anything which will give such a taint to their harbouring-places that they cannot feed, will do good. Soft-soap, tar, or, in fact, anything oily, greasy, or sticky that can be well rubbed on, and which, by adhering for a time, will choke all the Aphides that it touches, will be of use. In the case of an orchard so badly infested that the owner had begun to clear the trees, an application of coal-tar, well rubbed into the infested spots with a hard brush, was tried and succeeded well. The trees were cured of the attack and became healthy. Another observer mentions that his trees on which this was tried were injured and some killed. Probably this different result was from the state of the trees. An application that would be perfectly safe on the blight-tumours of old trees, would be very injurious on young bark that was still living and in an active state.

For washes, or mixtures to be laid on as paint, the following applications have been found of service ; but it should be observed that in the case of tobacco-water it is desirable to try what strength tender leafage will bear without injury :—

Take a quarter of a pound of tobacco, infuse it in half a gallon of hot water; when cool enough dip the infested shoots in it for a few seconds, or wash the infested parts in the liquor. Repeat this in a few days, if necessary, after which the plant may be washed with clean water. Then dissolve one pound of soft-soap and one gallon of lime in enough water to make it about the consistency of thick whitewash. Apply this with a painter's brush to the stem and all the branches that can be reached, and sift some lime on the ground.

An application is also recommended of half a peck of quick-lime, half a pound of flour of sulphur, quarter of a pound of lamp-black, mixed with boiling-water so as to form a thick paint; this to be applied warm. In winter, when the leaves are off, the branches and stems may be painted with this, *all loose bark being first removed*. It is very desirable to remove the soil from the bottom of the stem, down to the main roots, and paint that part also.

For special applications to nooks and crannies anything that is oily, soapy, or greasy will do good, but, as far as killing the insects is concerned, the thicker it is the better, so that it may fairly fill up the crevices in the bark, if possible, and not run off the Aphides till it has killed them by choking up their pores, but at the same time care should be exercised not to oil or grease *young* bark that may be hurt by the application.

"Machine oil" is *stated* to be a perfect remedy, and to do no harm to the bark. Pure whale oil is said to answer perfectly, put on as a paint on the stems and into the crevices. Neats'-foot oil is also advised. Probably any cheap oil will serve the purpose; the great point is to get it thoroughly into all the crannies with a brush, so as to come in contact with the plant-lice, and that many of these creatures may be rubbed out and killed in the operation. A simple lather of yellow soap, laid on with an old shaving-brush, does all that is needed sometimes without fear of hurting the plants, and soft-soap, well rubbed in, would probably be a very effective and lasting remedy.

It is also recommended, on good authority, that about the end of February the trunks and large branches should be scraped, excrescences cut off, and the whole well scrubbed with soap-suds, after which a good coating of lime and water is recommended. Probably the form of "whitewash" that has some "size" in it would be better than the simple lime and water, as the "size" makes it stick better, and thus it is more injurious to the Aphides.

Ammoniacal liquor, diluted with ten to twelve parts of water, will kill Aphides, but (for the reason of it varying in strength, as mentioned elsewhere), experiment should be made as to the quantity of water to be added to make it a safe application to such leafage as it may touch before it is used on a large scale. Apple trees that have the shoots and leaf-stalks infested may be well cleaned for a while by means of water sent from a garden engine with a flexible hose, or, where the water supply is laid on with a good pressure, the use of a hose with a spreader, throwing the water with considerable force, is better still. A strong and steady stream of water should be directed against every part; this will wash out the crevices, and knock off many of the Aphides, as well as help to keep up a healthy leafage. If the attack reappears, the treatment should be repeated.

For washes of soft-soap ; soft-soap and sulphur ; and gas-water and sulphur, with more or less soft-soap, according as the application is needed for a wash or a paint on the boughs, see references to "Washes" in Index.

A careful watch, and something done as soon *as the wool appears*, is what is wanted; but if the small tufts are left alone, as of no consequence, the insects will soon spread far and wide, and a thoroughly infested tree is an injury to a whole neighbourhood that ought not to be allowed.

With regard to the Woolly Aphis on Apple roots, doubts have been expressed whether it is of the same kind as that infesting the trees, but it appears now to be considered to be so, and, excepting in the matter of stopping passage down to the roots from an infested tree, this question does not affect means of prevention. Where Woolly Aphides are found on the roots it is advised by Dr. Asa Fitch to clear away the soil as much as possible from the infested roots, and pour strong soap-suds in sufficient quantity to soak into all the crannies or diseased spots, and either to remove the old soil and replace it with fresh, or to mix ashes with it.

Another observer recommends partially laying bare the roots, and following this up by the application of night-soil. Ammoniacal drainage from stables is said to cure the evil. As the root Aphides in all likelihood pass down from the trunk, it would probably be a great preventive to put a loose rope of hay soaked in tar round the tree, at its junction with the ground, placing the band so as to stop passage but not injure tender bark.

Besides the above applications, so many others are mentioned as being used with more or less success, it may be worth while to give the list in some kind of order.

We find it includes tar, kerosine, paraffin, turpentine (diluted), also resin (with an equal quantity of fish-oil, and put on warm); oils of various kinds; soaps of various kinds; ammoniacal liquor from gas-works, and ammoniacal animal fluid, especially drainings from stables; tobacco-water; paints of lime and soap; lime and sulphur; whitewash; oil and soot; and also plaster of grafting-clay to stop up chinks with the blight enclosed.

Of this vast collection of means of remedy, probably the most serviceable are thorough drenchings of some of the soap-washes applied by means of the garden-engine to the tree directly the attack is noticeable, accompanied by special applications of thick mixtures of soap, or of any kind known to be desirable to kill such of the blight insects as may have remained sheltered in crevices of the bark.—('Gard. Chron. and Ag. Gazette," 'Eighth Report of State Entomologist, Illinois,' Ed., &c.)

Apple Aphis. }
„ Green-fly. } *Aphis mali*, Fabr.

These Aphides infest the leaves, and are sometimes very injurious.

The following notes are taken from the excellent account given of the species by Mr. G. B. Buckton, F.R.S., in his 'Monograph of British Aphides,' vol. ii., that being the only life-history I am aware of giving information up to the present period, and to it I refer the reader for full details :—

"The black eggs of *A. mali* may be found deeply buried in the crevices of the bark, and these hatch as soon as the spring sap begins to swell off the buds. The young Aphides puncture the backs of the thick fleshy leaves, which pricking causes them to curve backwards from their points; and in this manner safe retreats are formed, and shelter from the effects of rain and hot sun."

" The winged insects abound most in July, when they spread their colonies so much that sometimes the vast orchards of Devonshire are wholly robbed of their fruit through shrivelling of the leaves. The bark of the trees sometimes is blackened by the glutinous secretion voided by these Aphides."

This species of Apple Aphis is stated to be very variable both in form and colour. Of the females that produce *living young*, the *wingless ones* (hatched from the egg first in the season, which may be called the mother-Aphides of the successive generations of the year), are globose and soft, larger than those born afterwards, of a dark slaty grey colour, mottled with green, with short dark grey horns and legs. The later viviparous broods are variable in colour, as green, yellowish, rusty red, &c. The *winged female* bearing *living young* (*viviparous*) has the head, horns, and body between the wings, black; abdomen green, with dots on each side; legs yellowish, with black knees and feet. The wings are long, and pale green at the base. The *wingless egg-laying* female is almost globose, of a brownish green colour, with a rusty stain on the head and part of the body next to it. The tail and rings next to it are very hairy.

In the case of this plant-louse, there are wingless males; whether there are winged ones also is not certain. The wingless kind is described as "exceedingly minute, perhaps one-eighth the size of the female" (of which the greatest length given is about the tenth of an inch), legs long, horns longer than the body, and sucker almost equal to it in length.

The early stages of this species of plant-louse much resemble each other in form; the pupa, however, has reddish wing-cases; also it is usually of a paler yellow in colour than the larva, and has three green stripes on the abdomen.—(' Mon. of Brit. Aphides,' vol. ii.)

PREVENTION AND REMEDIES.—" Where practicable, a syringing with tobacco-water made by pouring four gallons of hot water upon a pound of tobacco will be found efficacious. Anointing the branches with soft-soap or strong soap-suds kills all bark-pests, and the alkali has been said so far to act on the sap passing to the leaves that it sickens the Aphides, and causes them to fall and die on the ground."—(G. B. B.)

In the case of this Aphis which blackens the bark and gives the tree a sickly smell from its excretions, thorough and repeated washings that will clean the leaves and shoots, as well as knock off the Aphides, are particularly useful. Where shoots are still in the first stages of attack, before the leaves are ruined, good drenchings applied powerfully by means of the garden-engine (as recommended in the case of American blight) are useful for this purpose, and may be of water, or of any of the washes mentioned. Washes containing soap or anything that will adhere to the Aphis, instead of being repelled by its mealy coat, are the most useful.

It is desirable to cut off all infested shoots that are past hope of recovery, or can be spared, and destroy them at once, so as not to allow the Aphides on them to fly or otherwise get about.

The common Blue Titmouse is especially useful in destroying Aphides; and the Cole, Marsh, Long-tailed, and Great Titmouse; also the Lesser Spotted Woodpecker, the Creeper, the Nuthatch, and the Warbler, are stated to be serviceable in clearing insects from Apple-trees.—('Observations of Inj. Insects,' 'Mon. of Brit. Aphides,' &c.)

Apple-bark Beetle. *Scolytus (hæmorrhöus?)*, Megerle.

This Beetle is sometimes injurious in old orchards, in which it has been noticed in Lincolnshire, Yorkshire, Notts, Worcestershire, and Wicklow, Ireland.—(M. D.)

The female beetle forms a passage under the bark, along which she lays her eggs. These hatch into fleshy yellowish white maggots with ochreous heads and brown jaws, which gnaw each their own gallery, starting from the sides of the central passage and increasing in size with the increase of the grub that formed them.

The maggots in the case observed were hatched about the end of May, fed on the bark or occasionally on the sap-wood till the autumn, and came out of the infested branch as beetles in April of the following year.

These beetles are very small, only about half to two-thirds of a line in length. Under a magnifying glass this small *Scolytus* will be seen to be black, with the front and hinder margins of the body behind the head (thorax) rusty colour, the wing-cases rusty at the tip, and the horns on the head ochreous.

This Apple-tree *Scolytus* appears only to attack un-healthy trees, but when it does occur it has great powers of destruction, for a few females which were allowed to establish themselves on the stem of a tree (for the sake of experiment) laid so many eggs that the maggots from them destroyed the bark for nearly a foot in length.— ('Gard. Chron. and Ag. Gazette,' 1845.)

PREVENTION AND REMEDIES.—Removal of all dying or unhealthy branches in which the beetle may be found to have established itself should take place directly the attack is noticed, and these should be burnt at once, to prevent all possibility of the insects in them spreading to neighbouring trees. For habits and means of prevention of *Scolytus*, see Elm-bark Beetle (*Scolytus destructor*).

Codling Moth. *Carpocapsa pomonana*, Schmidberger.

The caterpillar of this Moth causes what are called "worm-eaten" Apples, which, falling a little before they are ripe, may be known by having a small discoloured spot with a hole in it on the lower side; from this a gnawed passage leads to the middle of the Apple, which is commonly nearly filled with dirt.

The method of attack consists in the moth (when the young Apples are beginning to form in the early summer) laying one egg in each fruit, usually in the eye of the Apple; from this the caterpillar or maggot hatches, and gnaws its way downwards, taking a direction so as not to hurt the core.

The caterpillar is about half an inch long, and slightly hairy; whitish, with a brown or black head and dark markings on the next ring, and about eight dots on the others; the food-canal shows as a dark line along the back. As it grows it continues its gallery towards the stem, or the lower side of the Apple, where it makes an opening through the rind, and thus is able to throw out the pellets of dirt which could not be got rid of by forcing

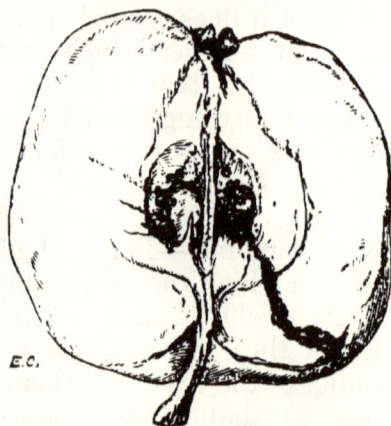

Apple injured by caterpillar of Codling Moth.

them upwards through its small entrance-burrow. After this opening is made it turns back to the middle of the Apple, and when nearly full grown pierces the core and feeds only on the pips; and as a result of this injury the Apple falls. After this the caterpillar leaves the fruit, crawls up a tree, and, when it has found a convenient crevice in the bark, gnaws a little more of it away so as to form a small chamber, where it spins a white web over itself.

Here in some cases (according to German observations) it turns to the chrysalis immediately, from which the moth comes out in a few days to begin a new attack on the fruit; or (as recorded in this country) it lives still as a caterpillar for several weeks, and then changes to the chrysalis, in which state it usually passes the winter; and from this the moth comes out in the following June.

The moth is about three-quarters of an inch in the spread of the fore wings. These have a light grey or ashy brown ground, with delicate streaks, and broader markings of a dark tint, giving a kind of damasked appearance; and at the hinder corner is a large spot of a brownish red or gold-colour, with paler markings on it, and a border of coppery or golden colour around it. The hinder wings are blackish.—('Ent. Mag.,' 'Illus. Brit. Ent.,' 'Naturgeschichte der Schad. Insecten.')

PREVENTION AND REMEDIES.—The infested Apples drop before they are fully ripe, but (as it has been observed that the caterpillars usually leave the fruit immediately it has fallen) removing these Apples does little good. If, however, where attack is prevalent, a slight shake was given to the tree, many of the "worm-eaten" Apples would drop; and clearing these away directly, before the caterpillars escaped from them, would much lessen the amount of future attack.

Looking at the habit of these caterpillars of crawling up a tree as soon as they have left the fruit, it is probable that throwing a shovelful of anything that they would not cross round the stems of the trees as soon as the worm-eaten Apples began to fall from the boughs, would do a deal of good. They probably could not cross gas-lime, or again, ashes or sand that had been well sprinkled with spirits of tar or with paraffin would most likely answer well. A band of hay or straw, well tarred and laid on the ground round the trunk of the tree to be protected, is a sure defence as long as the tar is wet, and, to keep it moist and sticky for a longer time, the addition of fish-oil, in the proportion of one ounce of oil to three ounces of Archangel tar, has been found suitable;

if the tar be moderately liquid, the addition of oil in the proportion of a quarter instead of a third would answer. For mixing, the tar and oil may be placed over a slow fire, but care should be taken not to allow them to become hot, as a very slight degree of heat will serve for all that is needed, and, if heated much, the mixture will dry more readily, or, if allowed to boil, will become hard and brittle on cooling down.

A band of soft-soap laid thickly a few inches wide just above the ground would most likely keep off attack, as few caterpillars would like to crawl over it. This kind of protection does good not only by keeping the trees clear, but also by keeping the caterpillars exposed for such a much longer time, that the birds are able to help us materially by picking up stragglers.

As the caterpillar harbours in crevices, it would be of service in this case, as well as with other Apple-pests, to reduce the number of these sheltering-places as far as possible, by dressing off loose bark and rough broken stumps of boughs, stopping up any particularly bad chinks, and—generally—keeping the trees in healthy and properly-tended condition.

It has been suggested that fires of weeds, such as would cause smoke (not flame), lighted near the trees at the time of egg-laying of the moths, would be of use; but probably the arrangement for "trapping" the caterpillars, mentioned by Prof. Westwood ('Gard. Chron.,' 1879) as practised by Mr. Wade in Tasmania, would do more good.

This consists in fastening a bandage of sacking or similar material round the stem of the tree from one to three feet from the ground, and it is advised that it should be put also round the branches. The grubs when seeking shelter "on coming to one of these traps seek no farther," and may thus be readily destroyed.

Lackey Moth. *Bombyx* (*Clisiocampa*) *neustria*, Curtis.

1, Cluster of eggs; 2, caterpillar; 3, moth.

The caterpillars of the Lackey Moth are injurious to Oak, Elm, Birch, &c., but are especially pests when they attack the Apple.

The eggs are to be found in winter and spring, laid on naked twigs, in compact spirally-arranged rings about half an inch long (see fig.). From these eggs small black hairy caterpillars hatch about the beginning of May, and immediately spin a web over themselves, which they enlarge from time to time as needed for their accommodation. In these webs they live in companies of from fifty to two hundred, and from them the caterpillars go out to feed on the leaves, returning for shelter in wet weather or at night. When alarmed, they all let themselves down by threads, either to the ground, or else after hanging in the air till the alarm is past they go up again by their threads to the tree.

When full fed, which is about midsummer, they are an inch and a half in length, and hairy ; of a bluish grey colour, marked with two black eye-like spots on the head, two black spots with a scarlet space between them on the next ring, and three scarlet stripes on each

side and a white one on the back, all bordered with
black along the rest of the caterpillar. At this stage
the caterpillars no longer live in companies, but each
finds some sheltered spot, between leaves, in hedges,
beneath the bars of railings, under roofs of sheds, or
even on the top of walls, where it spins a sulphur-
coloured silken cocoon, mixed with sulphur-coloured
powder and with hairs from the skin woven into it, from
which the moths hatch in July.

The moths are variable in colouring, mostly with
rusty fox or ochrey markings, but some have the fore
wings of a red-brown, with two pale ochreous streaks;
others yellowish, with dark brown bars; and others are
variously tinted: the hinder wings are reddish brown.

It is stated that the moths, and especially the females,
seldom fly, but remain concealed by day under leaves and
in long grass, and come out at night.

The caterpillars seldom do the enormous quantity of
mischief with us that they are noted as causing in
France, where, according to the old law, it was compulsory
on proprietors to have the webs on the shoots cut off with
shears and destroyed, in consequence of the ravages of
the caterpillars (if left unchecked) ruining the Apple-
leafage over an extent of miles of country; nevertheless
their attacks are often the cause of much loss in this
country, and need attention.—('Brit. Moths,' 'Illus.
Brit. Ent.,' and 'Gard. Chron.')

PREVENTION AND REMEDIES.—Some good may be done
by looking for the rings of eggs on the shoots, cutting
these off and destroying them; also by destroying any
yellow silken cocoons that may be found about the trees;
but these methods are tedious, and, though they are of
use where just a few trees can be carefully tended, are of
little service in orchard treatment.

A far better way is to watch for the webs, and, as soon
as they are seen, to carry out the old French method and
cut the shoots through with a pair of nippers and destroy
them. It is well for one person to cut, and another to

hold a pail below for the web and all the caterpillars (which on the first alarm would throw themselves down by their threads) to fall into. The pail should have a few inches depth of water in it, or mud thick enough to prevent the caterpillars from escaping.

A less troublesome but less complete method is to shake the boughs, or strike them smartly, so as to make the caterpillars drop, and sweep those that dangle by their threads in the air, down with the hand. These may be trampled on, or gas-lime, quick-lime, or anything that will kill them, may be thrown on them; but it should be done *at once*.

As the moths harbour under leaves and long grass, a properly kept state of undergrowth in orchards, free from overwhelming weeds and rank herbage, is of service in preventing attack.

Much more attention to this matter is needed relatively to keeping down Apple-pests than is commonly supposed. The dark, damp, confined air of the neglected over-crowded orchard fosters all kinds of insect-pests, and as no grass cut in such circumstances would dry, it is often left for rough feeding, or an occasional "skirming" of what is too long to remain uncut, and thus "pests" have possession; whilst where the trees stand apart, as they should, there is sunshine and fresh air to cause ripened growth, and lighten up the dark nooks that insects hide in; the grass can be properly pastured and attended to, and also the small birds have fuller access to do their work as insect-clearers.

Small Ermine Moth.   *Yponomeuta padella*, Stephens.

The caterpillars of this Moth are particularly destructive to the leafage of Apple-trees and also of Hawthorn hedges. The female lays her eggs in roundish patches on the small twigs, and covers these patches with a kind of strong gum, which is yellow at first, but gradually changes to a dark brown, so as not to be easily distinguishable from the brown twigs. The eggs may be found hatched by the beginning of October, but the caterpillars (which are then little yellow creatures with black heads, and only about half a line long) remain sheltered under the patch of gum during the winter, and do not come out till the leaves begin to unfold in spring.

Small Ermine Apple Moth, caterpillars, and cocoons.

Then it appears (see 'Trans. Ent. Soc. Lond.,' vol. i., p. 22) that they burrow into the young leaves and feed on the soft matter within, until they are strong enough to eat straightforward at the whole leaf, when they come out from their minings and thus make their appearance suddenly in large numbers where none have been noticeable just before.

The shoots may be seen covered with caterpillars early in the summer, and these continue feeding on the leaves and spinning webs in which they live together in large companies, until, in severe attacks, the hedge or tree infested is stripped of its foliage, and left hung over with a kind of sheeting of the dirty ragged remains of their deserted webs.

When full fed each caterpillar spins a light cocoon in which it changes to the chrysalis inside the general web.

The moths, which come out towards the end of June, are about three-quarters of an inch in expanse. The fore wings are usually livid or whitish, dotted with black; the hind wings livid or lead-colour; but they are so

variable as to be divided in some of our lists into distinct
species. The kind figured opposite (see 'Gard. Chron.
and Ag. Gazette,' 1849), characterised by the white
ground of the fore wings, thick cocoons, and by feeding
especially on the Apple, is sometimes distinguished as
Y. malivorella.—('Illus. Brit. Ent.,' R. H. L. in 'Trans.
Ent. Soc.,' vol. i., &c.)

PREVENTION AND REMEDIES.—As the caterpillars of this
moth turn to chrysalids in cocoons in their large nests
or masses of web, the simplest method of prevention is
to cut these webs off and destroy them.

Something towards getting rid of the caterpillars at
an earlier stage may be done by shaking the trees or
striking the boughs, so as to make the grubs fall; but,
from their habit of letting themselves down by their
threads on any disturbance and then going up them
again, as there is sometimes much confused web
about the leaves, it takes some trouble to ensure that
they do not return.

If they can be shaken in parties from large nests or
webs, they may be caught in a pail held just below, like
those of the Lackey Moth; but if they are scattered, care
should be taken to sweep down with the hand all that are
hanging at the end of their threads before they have
time to go back. This will prevent one method of return
to the boughs, and if anything they cannot cross is put
on the ground round the stem of the tree (as a ring of
tarred hay, a good thick painting of fish-oil soft-soap, or
anything else that may be known to be avoided by
caterpillars), then all means of getting back again
appears to be cut off. It is desirable, however, to
trample on the ground beneath a shaken tree, or throw
quick-lime, to get rid of the wanderers.

It has been observed that the whole brood of moths
usually hatch from the chrysalis at the same time,
when their light colour makes them easily seen, and they
are sluggish by day; it has therefore been found useful to
spread a sheet under the trees, and by beating or shaking

the boughs make the moths fall into the sheet and destroy them.

Fish-oil soft-soap, being procurable at a price of ten to twelve shillings per firkin of sixty pounds, is an application that can be used at remunerative rates, and, as in many other cases of insect-attack, good drenchings of it would probably be of service in clearing the trees of moths when they are hatching out of the chrysalids; and it would help to disperse the caterpillars if syringed on to, or into, their nests.

Mussel Scale. *Aspidiotus conchiformis*, Curtis.

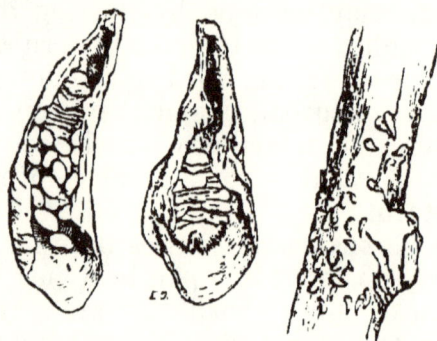

Apple Scale; with female; female and eggs, magnified.

The Mussel Scales are so named from their resemblance to very small mussel-shells. They are to be found both on Apple and Pear-trees, but chiefly on the former, where they occur on the more tender bark of the trunk, as well as on the branches, in such large numbers as to cause serious injury. Some kinds of Apple-trees are more infested than others, and the "Wellington" is especially subject to attack.

The Scales are about the eighth of an inch long, dark brown, slightly curved and rounded at one end, much smaller and of a rusty colour at the other, and wrinkled across. They adhere firmly to the bark, and on lifting full-grown specimens the females will be found inside

the smaller end of the Scale (sheltered by it, *not* fastened
to it), the larger end of the Scale being filled with fifty
or more white oval-shaped eggs. The young Scale-
insects that hatch from these eggs are very small, flat,
and white; furnished with eyes, horns, six legs, and a
sucker. These run about with great activity for a few
days, but after a while fix themselves and begin to grow,
and gradually change in appearance and turn to pupæ.
The female resembles a fat fleshy maggot of a greenish
colour, globular, somewhat flattened, and with lines
across showing a division into rings, but without
articulated limbs; after depositing her eggs she dies,
and may be found shrivelled inside the Scale. The
male does not appear to have been yet observed.—
('Gard. Chron.,' 'Prak. Insecten-Kunde,' &c.)

Prevention and Remedies.—Scale may be removed
at any time of the year, but the best season for destroying
it or applying dressings is in spring, so as to clear it
away before the young insects which creep out in May
from under the old dead shells have appeared, to begin
the new attack.

It may be removed by thoroughly moistening the
surface of the infested bark with lathers of any kind of
soap (or any dressing that may be preferred), and then
scraping the surface with a blunt knife, or rubbing it
with pieces of coarse canvas, or well brushing it, so as to
clear off the Scale without hurting the bark.

Scraping with a blunt knife is a good plan, as in this
way the Scales, moss, and everything on the surface are
mixed up in a plaster with the soapy lather, and got
thoroughly rid of together; if brushing is preferred,
good drenchings of soap and water, or of dressings
poisonous to the Scale, should be given in addition to
the first thorough moistening, so as to wash down or kill
all that may have only been disturbed or be lodged in
crevices.

Soft-soap or common coarse household soap are
useful for this purpose, and the following recipes for

dressings are mentioned as having been found serviceable, and might be varied in proportion of the ingredients
as thought fit.

One ounce of soft-soap, one pound of tobacco-paper,
and four handfuls of sulphur to one gallon of water; this
is to be applied with a painter's brush, taking care to
rub thoroughly; use plenty of the liquid, and flood every
part of the tree. Three applications in this way are
stated to have been always found to be a complete
cure.

As a means of clearing the Scale out of crevices, it is
advised to scrub the trees well at the proper season
(that is, during April or early in May) with soft-soap and
water, and then brush them over with the following
mixture:—Two pounds of soft-soap and one pound of
flour of sulphur, well mixed in about fourteen gallons of
water.

The following mixture has been found serviceable in
destroying Scale-insects, Thrips, and other plant-
vermin:—One hogshead of lime-water (use half a bushel
of lime to this quantity of water); add four pounds of
flour of sulphur, six quarts of tobacco-water, and four
pounds of soft-soap. This mixture is to be well stirred
and incorporated together, and applied by dipping the
infested boughs or by syringing. The composition may
be allowed to dry and remain on for about a week or ten
days, when it may be washed off with clear water.

It is also said to answer to get some tenacious clay,
dilute it with water to about the consistency of paint,
and to every gallon of this add half a pound of sulphur;
mix them well, and paint the trees all over. It is
advised to apply two dressings of this, allowing the
first to be thoroughly dry before the second is put on.
It requires a fortnight to kill the Scale by this application, and when the clay drops off it will bring the
Scale with it.

For Apple trees on walls, it is advised to un-nail the
trees, prune the young wood back as far as can properly
be done (and carefully burn the cuttings), and then wash

the wall with cement-water and the branches with any
application known to answer.

Oil has been advised, in the proportion of two parts of
cold boiled linseed-oil with one of the same oil raw, as a
sure means of destroying Scale; but the application of
oil where it can soak into the bark of trees has been
found in various cases so very prejudicial to its health
that it is not to be recommended generally, and is
merely alluded to on account of the great risk of injury
from its use.—('Gard. Chron.,' &c.)

Apple-blossom Weevil.   *Anthonomus pomorum*, Curtis.

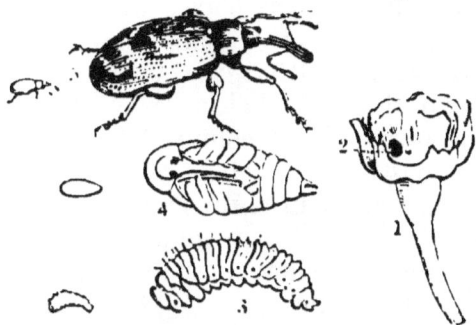

1 and 2, Apple-bud pierced by Weevil. 3, maggot; 4, pupa; 5, Weevil;
all magnified, with figures showing nat. size.

These Beetles attack the flower-buds of the Apple just
before they begin to expand, and in backward seasons
they sometimes do much harm in the cider counties;
such was the case in 1816, 1831, 1832, and 1838.

The female lays only one egg in each flower-bud; this
she inserts by boring a hole for it with the jaws which
are placed at the tip of her long curved proboscis; she
then lays an egg in the hole, and with the help of her
proboscis she closes the opening; she then goes on to
another bud, and may continue egg-laying for two or
three weeks; but the date and amount of attack depend
much on variation of the season influenced by the
weather, for the buds must be formed before the eggs

can be laid, and immediately the petals begin to unfold egg-laying ceases. Hatching may take place from the beginning to the end of April; if the weather is warm the eggs hatch in about six days. Meanwhile the bud grows and the petals are of their usual colour, but presently, instead of opening, they wither, and inside, in place of the stamens and germ, which have been eaten away, will be found a curved fleshy whitish wrinkled maggot, legless, with a few hairs, and a black horny head. This maggot turns to a pupa of a rusty brown colour in the hollow chamber in the bud, and in about a month from the time when the eggs were laid the Weevils develop from the pupæ, and disperse over the tree, where they feed on the leaves during the summer.

These beetles are of the shape figured above, of a reddish brown colour, with three indistinct stripes of a paler colour on the body behind the head; the wing-cases have a large pitchy-coloured patch, with a pale oblique stripe on it, and two ochreous spots towards the tip.

They pass the winter in chinks and crannies, or under loose pieces of the bark, or under clods of earth or stones, and come out when the flower-buds are swelling in spring, when the males may be seen flying round the trees, and the females generally crawling on the branches. —(John Curtis, in ' Gard. Chron.,' 1844.)

PREVENTION AND REMEDIES.—Much good can be done by clearing away all rubbish round the trees that may serve for shelter during the winter, and also by removing rough useless bark, and generally keeping the stems and branches of the trees in a well-tended condition.

Where the bark is clean and in good order there will be few hiding-places on the trees, and, as the female Weevils seldom fly, a large proportion of attack may be prevented by putting tarred bandages, or throwing something that they will not cross, on the ground round the stems, and thus preventing the females that have wintered under rubbish near the tree crawling up it in spring to lay their eggs.

The Weevils fall to the ground on being alarmed, and at egg-laying time many might be shaken down from the trees on to cloths spread below, and thus got rid of at an expense which would certainly be remunerative in garden cultivation, and worth a trial for orchard ground in cases where seriously bad attack was known to be going on.

It will be noticed that the Weevil lays no more eggs after the flower-buds begin to expand, so that in fine sunshiny weather the time of egg-laying is much shortened; also she lays on calm days, sheltering herself from wind or frost; and following up these habits it would appear that (in addition to other precautions) trees so placed and managed as to have plenty of sunlight and air around them and amongst the boughs are less likely to suffer, than where close-growing trees (even at the flowering time) keep sunlight and air from circulating properly, and where, though the buds on the upper parts of the trees expand in their due season, there are many others coming on slowly on the imperfectly-ripened wood beneath still available for the Weevil.

Many of the applications noted under the head of American blight would be of service in preventing attack of this Weevil.

CHERRY.

Cherry Aphis. $\begin{cases} \textit{Aphis cerasi}, \text{ Fab.} \\ \textit{Myzus cerasi}, \text{ Buckton.} \end{cases}$

This Aphis, which is often known as "Black Fly," is found in such enormous quantities on Cherry-trees in the early summer as at times to cause great damage. This arises chiefly from loss of the sap sucked away by the Aphides, but partly from the pores of the leaves being choked by the sweet sticky fluid ejected by the insects.

The females, both winged and wingless, that produce living young are black; the egg-bearing females are much smaller, and of a dark or ochreous-brown; and the male has the head and the body between the wings brownish black, and the abdomen of an ochreous-yellow, with brown transverse bars and spots.

The pupæ are olive-green, with yellow wing-cases.— ('Mon. of Brit. Aphides,' &c.)

PREVENTION AND REMEDIES.—Picking off the infested shoots and destroying them with the Black-fly adhering relieves the trees, and diminishes the amount of the coming attack.—(A. S.)

The following application is stated to answer well:— Mix some tenacious clay with water, so as to form a thin puddle. Then loosen the infested shoots (in case of the tree being on a wall), dip them in the mixture, and leave them to dry. After the imprisoned Black-flies have died the mixture may be washed off. This application should be made in tolerably settled weather, when there is neither violent rain to wash the mud off nor hot sunshine to crack it.—('Gard. Chron. and Ag. Gazette.')

Probably a little Paris green or hellebore infused in the water would make this application much more certain.

Probably thorough drenchings with some of the regular Hop-washes used to destroy the Green-fly would be about the best treatment also for this Aphis; a good stream of water driven at the tree is of service both in knocking Aphides from the shoots and in cleaning the coatings of dirt off the leaves; and it is also desirable to add something of a sticky nature that may adhere in some degree to each Aphis, and may be fairly reckoned on as likely to clog them together in the mass.

For this reason soft-soap is always a useful ingredient in an Aphis-wash, although tobacco-water, bitter aloes, and many other additions are serviceable, as specified under the heads Hop Aphis, Plum Aphis, &c.

The following application, in which a certain amount of sulphur is dissolved in the wash, would probably be highly serviceable :—Take four ounces of sulphuret of lime (which may either be purchased or produced by boiling the sulphur with lime), and two ounces of soft-soap to each gallon of hot water; the soap and sulphuret to be well mixed, and the hot water gradually added, the mixture being stirred during the time. A uniform fluid is thus obtained, which may be used when cool enough to bear the hand in it, and is serviceable for destroying many pests either by dipping infested twigs in it or by brushing it well into all nooks where the insects may have settled themselves, whether on the branches or in crannies of the walls.

GOOSEBERRY.

Magpie Moth. *Abraxas grossulariata*, Stephens.

Moth, and larva.

The caterpillars of this Moth are not so injurious as those of the Gooseberry Sawfly, but they occasionally occur in sufficient numbers to strip the bushes of their leaves. They frequent the Gooseberry and the Black and Red Currant, and also the Common Sloe, or Blackthorn.

The egg—one or more—is laid on the leaves towards the end of summer; the caterpillars hatch in September, and feed for a short time; and then either fall to the ground with the fall of the leaves in autumn, and remain sheltered amongst them for the winter (E. L. T.), or they spin the edges of a leaf together, which they have previously fastened by threads to the bough, and inside this protection remain until the return of spring.—(E. N.)

When the new leafage unfolds, the caterpillars come out and feed till May or the beginning of June, when they change to chrysalids.

The caterpillar is one of the kind known as "loopers," from the peculiar looped shape it draws itself up into when alarmed (see fig.); the head is black; body creamcoloured, with a reddish orange stripe along the sides;

the whole of the second ring, and the under side of the third and fourth, and of the four nearest the tail, are also reddish orange. A row of large irregular black spots runs along the middle of the back.

When full fed it spins a light transparent cocoon attached to twigs, or palings, or in crevices of walls; and in this it changes to a chrysalis, yellow at first, but afterwards shining black, with orange-coloured rings.

The moth is very variable in appearance; commonly it has a black head, yellow body between the wings, with a large black spot in the middle; the abdomen also yellow, with five rows of black spots. The wings are white, spotted with black, and the fore wings have a yellow blotch at the base and a yellow band across them. There are, however, almost endless varieties of markings, from black of different shades, to white; some have the upper half of the wing white and the lower black, or the reverse; some have the ground colour of the wing (instead of merely a band) yellow; and in some cases the hinder wings are striped with black.

The moths appear about midsummer or rather later. —('Hist. of Brit. Moths,' &c.)

PREVENTION AND REMEDIES. — The best method of preventing attack from these caterpillars in the spring is to destroy them whilst they are torpid during the winter. There is a difference of opinion as to whether the caterpillars winter beneath the fallen leaves on the ground or each in its folded leaf-cradle hung from the bough. However this may be, they may easily be got rid of by gathering the fallen leaves together from under the bushes, scraping up just a film of the surface-soil with them (so as to ensure none of the grubs being left behind), and at the same time casting a glance over the bush and picking off any hanging leaves that may be seen.

This is best done shortly after the leaves have fallen, but before there has been time for them to be dispersed by the wind, or for the caterpillars to be hidden in surface-crevices. The clearing of the boughs is most conveniently done at pruning-time, but, if this does not take place till late in the winter, it is probable that many of the leaf-cradles will have fallen, and the tenants thus escape being destroyed.

Thorough cultivation is a good preventive for this attack; where the bushes are properly pruned, and consequently carefully examined during the winter, there can be very little harbourage left on the boughs for the caterpillar; and a good forking beneath the bushes, with an addition of manure, especially of the rich sorts applied in Gooseberry-growing districts, cannot fail to much diminish the number of caterpillars sheltered on the ground. A ring of ashes sprinkled with tar, or of gas-lime, put towards March or April round each Gooseberry-stem (at a few inches from it), would keep any caterpillars from being able to crawl up it.

When caterpillars are first found to be attacking the bushes (as they do not throw themselves down readily on disturbance), many may be got rid of by a man with scissors in one hand and a jug or pail (with some mixture in it that the grubs cannot escape out of) in the other, snipping off the infested leaves into the vessel. —(J. C.)

Dusting the bushes with sulphur, quick-lime, soot, or any other of the usual dressings, will do good, and probably be very much more effective if applied in the dew of the morning than later in the day.

Syringing the leaves with Gishurst compound has been found very serviceable (F. A.); and probably lime-water would be of use, or alum, used as a weak solution lightly syringed over the leaves, as recommended for "Gooseberry caterpillar" by J. M'L. Syringing with soap-suds or a solution of soft-soap is always a means of lessening insect-attack, and would be useful also for clearing off Green-fly, which is often injurious to Gooseberry-bushes.—'Notes of Observation of Inj. Insects,' &c.)

Gooseberry and Currant Sawfly. { *Nematus ribesii*, Curtis. *Tenthredo grossulariæ*.

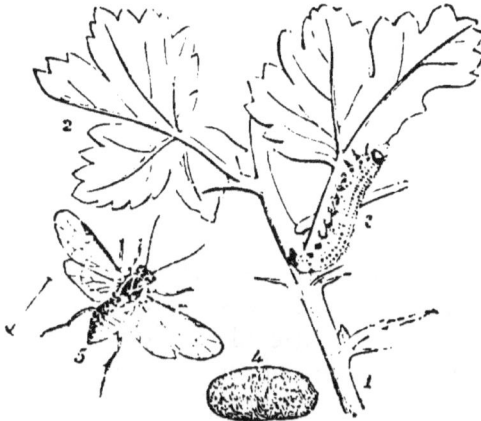

Gooseberry Sawfly; larva and pupa.

We are only too well acquainted with the havoc caused by this grub to the leaves of our Gooseberry bushes, as the attacks occur throughout the country yearly, and to a serious extent.

The female sawfly appears about April, and lays her eggs on or beside the largest veins beneath the Gooseberry leaves. The grubs hatch in about a week, sometimes less, rarely more, and begin to feed at once by eating small round holes in the soft part of the leaf. As many as sixty or seventy grubs are recorded as having been found on one leaf, and the attack of the grubs on the under surface may be discovered by the appearance of a number of small holes pierced through from below.

The grubs feed on the leaves on which they were hatched until each leaf is stripped of all that is eatable, after which they disperse themselves over the bush, and, in a bad attack, eat away the whole of the leafage save the stalks and some of the hard veins.

These caterpillars may be easily known from those of the Gooseberry moth by having a larger number of

sucker-feet; they have a pair of sharp horny feet on
each of the three segments next to the head; the fourth
segment is footless, but the following *six* segments are
each furnished with "*sucker-feet*," or "prolegs," like
short fleshy legs, and there is a similar pair at the end
of the tail, known as the "caudal proleg," making twenty
feet in all.

These grubs are of a bluish green, with black head,
feet, tail, and also black spots on each segment, and
with a yellowish space just behind the head and another
just before the tail.   When full-grown, and after changing the skin for the last time, these yellow patches still
remain, but they are otherwise of a delicate pale green,
with sometimes two little black dots on the head, and
are about three-quarters of an inch in length.   After the
operation of casting the skin they rest awhile, and then
crawl down the stem of the bush or drop from a bough,
and at once begin to bury themselves.   When deep
enough, which may be from two to eight inches, according to the nature of the soil, they form a cocoon of a
gummy secretion, in which they turn to chrysalids.
This takes place in about three weeks during summer;
in the case of the late broods the grub remains unchanged in the cocoon during winter, and does not turn
to the chrysalis till spring, in time for the Gooseberry
sawfly to make its appearance as the Gooseberry and
Currant bushes are coming into leaf.

The sawfly is of the shape figured above, and of the
length marked by the straight line; the head and body
between the wings are ochre-colour or yellow, variously
marked with black; abdomen yellow or orange; legs
yellow, with brown or black tips to the feet and hinder
shanks; horns brown or black.   The four wings are
transparent and iridescent.—('Gard. Chron.,' &c.)

PREVENTION AND REMEDIES.—An excellent and effectual
method of preventing attack in the coming season is to
remove the soil from beneath the bushes to the depth of
a few inches early in spring, and give a good sprinkling

of lime ; by this means the caterpillars which winter in the ground are completely cleared away.—(J. M'K.) A slightly different method, but thoroughly successful, is to remove the surface-soil below the bushes in winter, dig a deep hole, and bury the whole of the removed soil, cocoons and all (so deeply as to ensure having no further mischief from them), and replace the earth removed with manure and the soil dug from the hole.— (A. A.)

This complete removal of the soil with the cocoons is quite worth while wherever Gooseberry caterpillar is prevalent, and a layer of unslacked lime, well mixed with the soil as deep as the cocoons are, would be highly beneficial in case of the surface-soil not being removed. Gas-lime also would be of service, well sprinkled on the surface, if fresh, or lightly pricked into the surface-soil beneath the bushes after it had been aired for a few weeks, taking care not to lay it against the stem. Drenching the soil with urine, taken from the tanks of the cattle-folds, is also a remedy.

When the caterpillars appear on the bushes, it is of great importance to attend to them at once ; whilst still very young two or three dozen may be found on one leaf and got rid of together, which in a few days would have spread themselves over the bush. This early stage of attack may be known by the leaves appearing as if riddled with dust-shot. At a later stage thorough hand-picking, or shaking the bushes so as to make the caterpillars fall, is of service. If the caterpillars are allowed to drop on the ground, they should be crushed with the foot, or with the back of the spade, but a surer plan, with little more trouble, is to spread cloths or put some tarred boards under the trees, and thus collect and kill them.

It has been found that syringing with water of a warmth just bearable to the hand is a good means of clearing the caterpillars (experimenting first as to what heat the foliage will bear, as the tender leafage may be discoloured or killed by a temperature that will not

cause the slightest injury to the leaves a few weeks later).
—(J. M'K.)

The following mixture has been recommended :—Three
gallons of warm soap-suds, half a pound of soda, half a
pound of salt, and a handful of soot; the bushes to be
syringed on a still day when the sun is off them. Half
an hour after the application the plants should have
clean water dashed over them. It is stated that this
mixture does not injure either the young fruit or leaves.
—('Gard. Chron.') Soap-suds syringed on the bushes
have been found a useful remedy.—(R. S.)

Syringing with tobacco-water has been found an
effectual cure, if applied in time (P. L.) ; and a solution
of alum and water, applied by a watering-can with a
rose slightly over the bushes, has been found to be a
good as well as cheap cure.—(J. M'L.) Lime-water is
also said to be serviceable. Quick-lime also does good as
a dusting over the bush, but the use of sulphur is
particularly to be recommended. Flour of sulphur "is
easily applied by dusting it over the bushes with a
pepper-box while they are under the morning dew, or, if
during dry weather, the bushes ought to be watered and
then dusted. It is only necessary to dust the lower part
of the bushes if taken in time."—(W. M'C.) This
treatment has been confirmed by further experiment,
showing the sulphur to be "as efficacious as hellebore
powder," without risk in the application.—(J. W. W.)

This matter is well worth attention, for, whilst on one
hand the effect of hellebore on the caterpillars is so
certain as to make it a generally adopted application, on
the other, the very serious illness following on partaking
of berries from which the powder has not been removed,
make it an unsafe treatment, and dangerous to advise.
—('Notes of Observations of Injurious Insects.')

NUT.

Nut Weevil. *Balæninus nucum*, Germar.

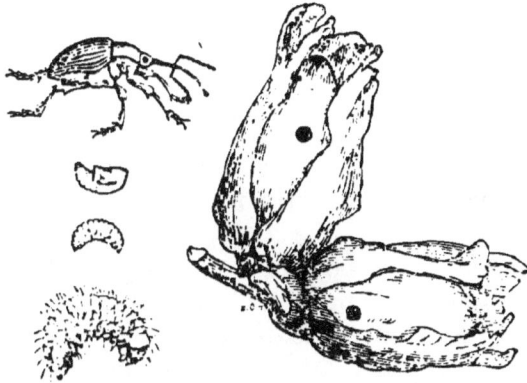

Nut Weevil; pupa; maggot, nat. size and magnified; pierced Filbert.

The fat whitish maggot found in Filberts and Hazel nuts is the larva of the Nut-weevil.

These beetles, or weevils, may be seen about the Nut bushes early in the summer, the females usually creeping along the twigs, the males often on the wing. Whilst the nuts are still young the female pierces a hole through the soft shell by means of the jaws with which she is furnished at the end of her long snout. In this hole she lays one egg, which hatches in about ten days. The maggot feeds inside the nut, consuming a large part of the kernel. When full-grown it is of the shape and size figured above, of an ochreous-white, with horny chestnut-coloured head, furnished with strong black jaws; without legs, but supplied with muscles inside the large transverse folds or wrinkles, which enable it to draw itself through the earth.

When full-fed the maggot eats a hole through the nut-shell, sometimes whilst the nut is on the bush, sometimes after it has fallen to the ground in the premature ripening which appears to follow on the maggot-attack. It then buries itself, forms a cell in the earth, and " frequently rests there during the winter, and only changes in the following spring or later " to a pupa or chrysalis of whitish colour, like the future weevil in shape, but lying quiescent with its limbs folded against it.

The weevils are of a tawny brown, and covered with short down ; the snout is of a bright chestnut colour. This snout or rostrum (see fig.) is furnished with a pair of elbowed horns or antennæ set half-way down it, and the wing-cases, which have ten lines of punctures on each, are variegated with reddish brown or ochreous marks.

The weevils may be found as early as May, at which time development from the chrysalis has been recorded, but it is stated that some of these beetles do not develop till July or August, and it is still open to observation whether some of these do not hybernate and appear with those newly out of the chrysalis in the following May.— (' Gard. Chron.,' ' Prak. Insecten-Kunde.'

Prevention and Remedies.—In this case the best means of prevention lie in the regular measures of good cultivation. It is stated the Filbert likes a hazel loam of some depth, " which should be dressed every year, as the Filbert requires a considerable quantity of manure." —(' Enc. of Gardening,' J. L.)

In Kent the Nut-grounds are " well-manured every other year with rags, shoddy, fish, or fur waste, and are always cultivated by hand, and kept scrupulously clean." —(C. W.)

This course of treatment, which involves stirring the surface-soil as well as additions, is suited to expose some of the chrysalids and bury others deeper, and is generally useful for insect prevention, but especially as regards the Nut-weevil, which (in instances observed) has been found to be so tender at the time of its transformations as to require eight or nine days to gain its colour and hardness, and also strength enough to force its way up through the ground. Looking at these points, it seems likely that if the chrysalids were buried a little beyond the natural depth many of the weevils from them would not be able to come up at all.

Where weevils are very abundant on the trees, it has been advised to beat them down, but this should not be done on a sunny day, or they will speedily take flight and escape; and (saving for treatment of a bush or two) probably the only way to carry out the plan of beating serviceably would be either to put tarred boards, or for one man to throw a sprinkling of quick-lime or gas-lime under the Filbert trees whilst his companion beats down the weevils.

It is desirable to remove all nuts that fall before their proper time, that the maggot inside may thus be carried away before it has bored its way out; and also, looking at the powers of flight of the weevils, it would be well not to have many Hazel-nut bushes in copses adjacent to Filbert ground.

PEAR.

Wood Leopard Moth, *Zeuzera æsculi*, Curtis.

1, Caterpillar; 2, chrysalis; 3, moth.

The caterpillars of this moth feed in the live wood of many kinds of trees. They are to be found in Pear, Apple, Plum, and Walnut; also in Ash, Beech, Birch, Elm, Holly, Lime, Oak, and others, besides Horse Chestnut (*Æsculus hippocastanum*), from which the moth takes its specific name, though not appropriately, as it rarely attacks this tree.

The eggs are laid during July, or later in the summer, in crevices of the bark, and on the branches as well as the trunk of the trees; these eggs are oval and salmon-coloured, and as many as three hundred have been seen laid by one moth. The caterpillars, which soon hatch, feed at first in the bark, but not long afterwards they make their way into the live wood, where they bore galleries rather wider than themselves, and as much as a foot in length. When full grown they are about an inch and a half long, whitish or pale yellow, with a black horny plate on the segment behind the head and.

another above the tail; the other segments are spotted with black, and the head is black or has two black spots. They feed wholly or at intervals until May or June (statements are made that they live for two years), and, when full-fed, they spin a web, or form a case of wood-dust, in which they change to an ochreous-brown, long, cylindrical chrysalis. This web is usually woven just inside the bark, near the entrance of the boring, so that when the time for development is come the chrysalis forces itself through the opening, and, by means of the fine prickles with which it is furnished along the back, it is held firmly in the web whilst the moth frees itself, and leaves the empty case projecting from the tree.

The moth is large and handsome, with a white head; the wings are somewhat transparent, and are white with black spots, the spots being darkest on the fore wings, which also have yellow veins. The body between the wings is white spotted with black, and the abdomen grey, or grey banded with black.

It is stated that the female moths appear somewhat later than the male, and may be found until the end of August.—('Illus. Brit. Ent.,' 'Gard. Chron. and Ag. Gazette.')

PREVENTION AND REMEDIES.—It has been suggested that fruit trees infested by this caterpillar sometimes bear better than those which are uninjured, but even if this is the case for a time, which is very much to be doubted, the health of the tree is gradually so much injured by the borings that sooner or later the fruit will suffer in consequence, and it is in every way desirable that the attacks should be prevented.

The caterpillars may be destroyed (like those of the Goat Moth) by drawing them out of their burrows with hooked wires, or by running a strong wire into the hole, and thus crushing the caterpillar within to death. If the wire, when withdrawn, is found to have wet whitish matter on it, such as would result from having crushed the larva, or again, if gnawed wood should have been

passed out of the burrow up to the time of the operation
and no more appear afterwards, it may be supposed the
creature is killed; otherwise the operation should be
repeated.—(W. S.)

Syringing is also of service in getting rid of these
caterpillars.   For this purpose a gutta-percha tube with
a sharp-pointed nozzle may be fitted to the syringe, and
thus, by placing the point of the nozzle well into the
hole, it may be filled with strong tobacco-water, soft-soap,
or any mixture that may be preferred, such as will make
the hole too unpleasant or poisonous for the grub to
remain in, even if it is not killed by the application.

The fumes of sulphur blown into the holes are also
very effective in destroying the caterpillars (M. D.) ; and
tobacco-smoke has been suggested for the same purpose.

This moth is preyed on by bats.

Jumping Plant-louse. Pear-sucker.	*Psylla pyri*, Schmidb. *Chermes pyri*, ,, *Psylla pyrisuga*, Först.

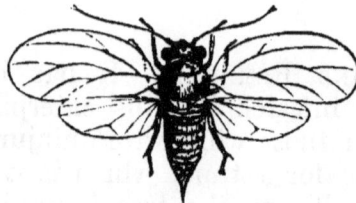

Pear-sucker, magnified.

In its early stages this little four-winged fly is injurious
to Pear-trees by sucking away the juices of the shoots,
and consequently hurting the leafage; and also by
voiding the sap in such great quantities as to defile
everything near it.

It is nearly allied to the Aphides, but more like a
Cicada or Frog-hopper in shape; the horns are peculiar,
from being furnished with two minute bristles at the tip;

and, from the formation of the hind legs, the fly has the power of leaping. It does not appear to have any regular name in this country, but, from the habit above mentioned, it is known in America as the "Jumping Plant-louse"; and, from its method of feeding, it is known in Germany as the "Pear-sucker."

The following notes of its life-history are from German observations :—

As soon as the Pear-trees begin to blossom in the spring, the male and female *Psylla* come out from their winter hiding-places, pairing takes place, and the females lay their eggs (which are of a yellow colour) either on the young shoots, the under side of the leaves, or some part of the flower, selecting for the purpose such parts as are covered with woolly hairs.

The larvæ commonly hatch in ten to fourteen days; they are of a dusky yellow, with lighter abdomen, head with whitish horns, and six legs. In the next stage they are darker, with dark wing-cases, and, from their broad flattish shape, very like small bugs or scales. In this state they crawl from the leaves and collect in great numbers on the shoots, and suck away the juices, which they discharge in such quantity that the presence of the *Psylla* may be known by the numbers of Ants and other insects attracted to the sweet excretion. When the pupa is matured it leaves the feeding-spot, fixes itself on a leaf, and, cracking the outer skin, the perfect insect comes out, leaving the pupa-skin empty. These Pear-suckers are about the size of a large Aphis, and are at first greenish, with red eyes; but gradually change to various shades of crimson, dusky red, and black. The four wings are snow-white, and the head is remarkably broad.—(Mainly from 'Prak. Insecten-Kunde.')

PREVENTION AND REMEDIES.—During the last season (1880) a note is given of this insect being most destructive to Pear-trees on walls, as well as standards and espaliers, ruining the leaves by extracting the sap, and also by the immense quantity of their gummy adhesive

·excretions. The treatment found serviceable was, first, syringing the trees with soft-soap, in the proportion of one pound of soft-soap to eight gallons of water, which helped to sicken the insect before using a further appli-·cation of Gishurst compound and tobacco-water mixed.

It is also noted as a successful method of prevention to put the proportion of soft-soap and water above mentioned into a large garden-engine and apply it all over the trees when there is not less than six degrees of frost, so that it becomes a sheet of ice as soon as put on. On thawing it clears off all adhering matter, as well as insect eggs, &c.—(J. W.)

This application would be well worth experimenting with relatively to clearing Aphis-eggs off all garden fruit-trees; I have experimented with it since receiving the ·above note, at a lower temperature than that named, without the slightest injury to the shoots or subsequent leafage, and, if found to answer for general application (by means of the contraction and expansion of the moisture, in freezing and thawing loosening and clearing the eggs out of inaccessible crannies), it would meet one of the points we need in Aphis prevention, as well as with regard to " Pear-suckers."

The Pear *Psylla* winters in protected nooks of the bark or similar sheltering places; therefore any measures to ·diminish the amount of these hiding-holes, or make them unpleasant to the insect, are of use; and, generally, the same applications and measures of prevention used for Aphides are serviceable here.

Pear Oyster Scale. *Aspidiotus ostreæformis*, Curtis.

These Scales are of the same nature as the Mussel Scale of the Apple, and injure the branches in the same way by sucking out the juices.

They are about the seventeenth of an inch across, and remarkably like oyster-shells (from which they take their name) in miniature. They are for the most part wrinkled, and slightly raised in the middle ; grey, with a lighter or white margin, and with a raised semi-transparent ochreous spot near the centre.

The female usually fixes herself so firmly to the branch by her sucker that she remains behind when the Scale (above mentioned) is lifted. She is much like a globular flattened somewhat heart-shaped maggot, yellowish white and shiny, with a few hairs, and yellow at the tail ; and is without wings or observable legs.

The male is ochre-colour, furnished with one pair of wings, six legs, and horns very nearly as long as itself; and has the abdomen suddenly narrowed to a long point. —(' Gard. Chron. and Ag. Gazette,' 1843.)

For further notes of life-history of *Aspidiotus* see "Apple Mussel Scale."

PREVENTION AND REMEDIES.—Syringing the infested trees with hot water and soft-soap during the winter clears them from the Scale, and prevents further attack. —(J. D.)

The same applications that are noted as serviceable for the Apple Scale would also be useful here, excepting that, as the Oyster Scales are very small and flat, probably rubbing with a cloth or brushing with soft-soap would be safer for the bark than scraping it.

Slug-worms.   (*Tenthredo (Selandria) cerasi,*
Larvæ of Pear Sawfly. (Curtis.

The Slug-worms feed on the upper surface of the leaves of the Pear and Cherry, clearing away the whole of the soft substance of the leaf, so that the veins and the skin of the lower side are all that remain ; they are also to be found on Plum, Hawthorn, and Sallow, from the middle of August until October.

For all practical purposes the Slug-worms may be known (when at their work of destruction) by their blackish or bottle-green colour, together with their

U

peculiar shape, and the covering of slime or moisture exuding from their skin, which gives them something the appearance of a slug, but still more that of a lump of wet black dirt fallen on the leaf and run together at one end. They appear also to be known by the sickening smell observable where many trees are infested.

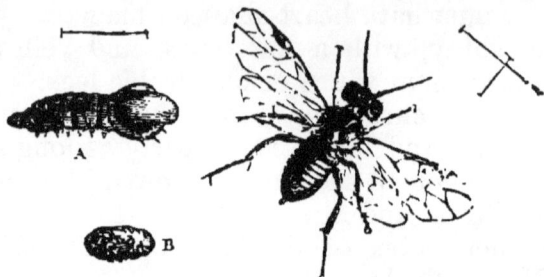

A. Slug-worm and Sawfly, magnified; lines showing nat. size.
B. Cocoon.

In respect to the scientific definition of the species there has been much difficulty, from some kinds of *Selandria* being nearly alike in their perfect state, and probably in appearance and habits whilst still " Slug-worms "; and consequently, whilst there are many synonyms for one species, there is also doubt whether two species are not given under one name. Therefore, for clearness, I give the following note entirely from the observation of John Curtis, attaching to it the name he applied of *Tenthredo cerasi*, and quote the description of colouring of the legs of the Sawfly in full, as this is one of the points requisite in determining the species.

The Sawflies appear in July, and deposit their eggs on or in the upper side of the leaf; these eggs are oval, and hatch in a few days. The larvæ are of the lumpy shape figured above, much the largest at the back of the head; they are furnished with ten pairs of feet, that is, one pair on each of the three segments next to the head, and a pair of sucker-feet on each of the other segments, excepting on the fourth from the head and the tail-segment, which are footless. When feeding, they keep the end of the tail a little turned up.

In four or five weeks these Slug-worms arrive at their full growth, which is about half an inch in length, cast their dark bottle-green skins, and appear as yellow or buff caterpillars, free from all shine, and transversely wrinkled, instead of being perfectly smooth. In the instance noted this happened at the beginning of October, and the caterpillars shortly after left the leaves and went down into the ground, where they spun an oval brown silken cocoon covered outside with earth, from which the Sawflies came up in July in the following year.

The female fly (figured above, magnified) is of a shining black, tinged with violet; the wings often stained with black, with dark nerves, and a dark brown mark (the stigma) along the fore edge.

"The four anterior legs are brownish ochre, and the others are more or less of that colour, but generally much darker; and the thighs, or at least the base, are pitch-colour."—(J. C., in 'Gard. Chron. and Ag. Gazette,' 1842.)

From the notice of "The Pear-tree Slug" published by the Entomological Society of Ontario it appears that the Sawflies are double-brooded in Canada. The flies appear in May; the eggs are deposited singly in little slits cut for them in the skin of the leaf by the ovipositor of the female, and these produce a brood, coming out in the perfect state in July; from which a second brood arises, which is full grown in September or October. These remain in the ground during the winter, and for the most part appear (as above mentioned) fully developed in the following May; but some remain in the ground unchanged till the following year.—('An. Rep. Ent. Soc. of Ontario,' published by the Legislative Assembly, 1874.)

PREVENTION AND REMEDIES.—An excellent method of prevention for the next year's brood is, to skim off the surface-soil beneath infested trees (as in the case of the Gooseberry Sawfly), and get rid of it so as to destroy the contents. The cocoons are stated to be at a depth of from one to three or four inches below the surface, according to the nature of the soil; and it would be desirable to ascertain this before beginning operations. A light stirring of the surface-soil and pointing in a good dressing of quick-lime would also be of service.

The Sawflies have been found to fall to the ground on the tree being shaken, and to remain for a short time motionless; consequently it would be a good plan to place boards covered with wet tar, or cloths, beneath the tree, and shake the flies down on them early in the morning or late in the evening (or at whatever time it was found they were collected on the leafage), taking care that they were destroyed before they could escape.

Dusting with caustic lime destroys the Slug-worms, but has to be repeated to make sure work, for it has been found that the larvæ at the first application can cast their skins with the obnoxious matter, but are not able to do so at the second dusting. The lime may be dredged on them from any kind of tin box pierced with holes; or, if something on a larger scale is needed, may be sifted through a sieve, or a common willow garden-basket may be very conveniently used for the purpose.

Tobacco-water will destroy them; and lime-water has also been found useful, in the proportion of a peck of lime to thirty gallons of water; it is noted that if two pounds of soft-soap are added, it will improve the mixture, and that the best times for syringing are before seven o'clock in the morning and after five in the evening.—(J. C., in 'Gard. Chron. and Ag. Gazette.')

Heavy syringings of the trees with strong soap-suds, applied by a powerful garden-engine, are very effective in getting rid of this pest.—(M. D.)

A solution of hellebore showered over the infested trees has been found effectual in clearing them of the Slug-worm.—('Notes of Observations of Inj. Insects,' &c.)

PLUM.

Plum Aphis. } *Aphis pruni*, Reaumur.
  „   Green-fly. }

"*Aphis pruni* is exceedingly destructive. Multiplication takes place by millions, and the insects close up the pores of the leaves by their tenacious excretions and the mealy exudations from their bodies. By the constant irritation of their rostra [suckers] the leaves roll up, and under this cover from the weather both the winged and apterous forms live overspread by the before-mentioned mealy powder, which probably to them is a protection."—(G. B. B.)

The wingless female bearing living young is of various tints, from green to slight olive-brown, with three faint green stripes on the abdomen, short olive-brown horns, and brown eyes; the winged female is apple-green; with head, horns, body between the wings, and feet, black.

The male is small, dingy ochreous, with the head, part of the body immediately behind it, some markings on the back, and the feet, umber-brown; the fore wings are large and broad; sometimes the insect is black. It is to be found in November in company with the egg-laying female, which is small, pale greenish yellow, and transparent; and usually shows the mature eggs within, which are ready for laying.

The variety of the Green-fly of the Hop (known as *Aphis humuli*, var. *malaheb*) is also common under the leaves of the Garden Plum in May and June.—('Mon. of Brit. Aphides.')

PREVENTION AND REMEDIES.—Amongst various washes recommended are tobacco-water, ammoniacal liquor, soap-suds, and soft-soap.

Where trees have to be treated on a large scale,

probably nothing would be more serviceable than some of the washes already noticed as in regular use, and found to answer well in the Hop-ground.

In one of these the proportion is of fourteen to twenty pounds of soft-soap to one hundred gallons of water, with the juice of a quarter of a pound of tobacco added, if thought desirable.—(C. W.)

Another, mixed on a large scale to be diluted for use, is of sixty pounds of soft-soap added to thirty-six gallons of water, with fourteen pounds of bitter aloes, or two of tobacco; the whole is boiled together, and, when applied, thirty-six gallons of water are added to every gallon of the mixture.

If these washes were applied as in the Hop-grounds, by means of a large garden-engine fitted with a gutta-percha tube and jet, or rose, or spray-syringe, so as to send the mixture under the leaves, as well as over and round the whole of the tree, and thus drench it down completely, the effect would be excellent.

A thorough application of this kind unites many good points; it knocks many of the Aphides off, it cleans the leaves of the accumulating dirt which is choking them, and makes the surface distasteful to the plant-lice for a while; also the soft-soap washes have the great advantage of sticking in some degree to the Aphides. When these insects (as noted above) are covered with a kind of mealy powder, many of the washes used simply run off them at once; and, unless the application sticks to them, so as to poison them, or is given so violently as to knock them from their position, the labour does little good.

On a small scale soap-suds are a good remedy. They may be used by syringing, or a lather of yellow soap may be applied to the infested parts with a brush, or the shoots may be dipped into the mixture.

Ammoniacal liquor diluted with ten to twelve parts of water will kill "Green-fly"; but this liquor (gas-water) varies in strength at different works, and the leaves also vary, according to their nature and age, in the power to bear such an application; therefore it is desirable to experiment where there is any doubt as to safe strength.

A recipe given for making tobacco-water is, to infuse half a pound of tobacco in a gallon of water, and dip infested shoots in the liquor, repeating the operation in a few days if necessary, after which the plants may be washed with clear water; a strong solution of tobacco has also been advised as a syringing for wall fruit-trees, washing (as mentioned) with clean water afterwards. Here, however, as with gas-water, some care is needed, for the strong solutions are apt to be injurious to the leafage as well as to the Green-fly; so that a mixture of soft-soap with a moderate quantity of tobacco-water will probably be a surer and safer remedy.

A decoction of quassia-chips and soap, or soft-soap, has been found serviceable both for syringing infested trees and also on a smaller scale for dipping shoots in. One recipe for the mixture is—One ounce of quassia boiled for ten minutes in a quart of water, and a piece of soft-soap the size of a small hen's egg then added. Quassia is a well-known "fly-poison," but, having found that sometimes flies which appeared to have been killed recovered afterwards, it suggests that the different amount of success from the use af this remedy may depend in part on the strength, but also on the Aphides being well washed down by syringing, or otherwise cleared from the shoots whilst they are still stupefied.

The enormous number of different applications recommended for destroying "Green-fly" make it impossible to enter into detail; but the same principle runs throughout—to get rid of the Aphides either by heavy syringings that fairly sweep them, or many of them, down; or by poisoning them, or coating them with sticky matter that will kill or injure them; or by making the surface of the leaves and shoots distasteful to them. With a foundation of soft-soap, or fish-oil soft-soap, and a small addition of anything poisonous or deterrent according to fancy, each cultivator may make for himself at a small cost a thoroughly serviceable wash.

Fumigation may be of service, even out of doors,. where, by means of matting supported on poles (or on anything else that will serve the purpose), the smoke can be confined, and thus be brought to bear on a badly-infested wall fruit-tree or an espalier; but, excepting in this way, though stated to have been found useful, it does not appear conveniently practicable.

A great deal might be done in the autumn of one year, to lessen the impending attack of the next, by cutting off all infested suckers and long succulent shoots, which are often left in neglected gardens to become a perfect head-quarters swarming with Green-fly. These should be carefully destroyed; probably a firm pressure under the foot, given to each shoot as it is cut off, is the surest way of getting rid of them; but if they can be burnt or thoroughly destroyed any other way *at once*, it will be all the better.

In almost all cases the oviparous female of the *Aphidinœ* (that is, of the tribe of the *Aphididœ* to which the Green-fly of the Plum, Hop, Turnip, Cabbage, Bean,. and of many other common plants, belong) lays her eggs. in late autumn, and dies. These eggs may be found attached to the twigs, or by the buds, sometimes almost singly, sometimes in clusters of several hundreds; and any measures by which infested shoots which are nursing attack in the autumn can be cleared are desirable; and,. though I am not aware that the plan has been tried, I believe a few heavy syringings of the trees with a strong wash about the beginning of October would probably be very serviceable in preventing egg-laying. These autumn egg-laying females being wingless, a good drenching,. even with mere water, such as would sweep them from the tree, would be of use.

Cold cannot be depended on to kill insect-eggs.— (' Gard. Chron. and Ag. Gazette,' ' Report of Observation. of Inj. Insects,' ' Mon. of Brit. Aphides,' Ed., &c.)

The Winter Moth. *Cheimatobia brumata*, Stephens.

Winter Moths, male; and female, showing abortive wings.

The caterpillars of this moth are injurious to almost all our fruit and forest trees. They feed on the young buds, and leaves of the Plum, Apple, Pear, Elm, Lime, Willow, Hawthorn, and many others, and occur at times in such great numbers as to cause a very serious amount of damage.

The moths (known as "Winter Moths," from the season of their appearance), come out about the end of October. During November and December the females, which have only abortive wings (see fig.), creep up the trees and lay their eggs on the leaf or flower buds, on the twigs, or in crevices of the bark.

The eggs are greenish at first, and gradually change to brown or red. They are very minute, and very numerous (a single moth laying as many as two hundred), and they hatch about the beginning of April. The newly-hatched caterpillars are only about as thick as a horse-hair, greyish in colour, and may be seen swinging in the air at the end of their threads; when full-grown they are half an inch long, of a yellowish green, with pale green head, black or blue line down the middle of the back, and whitish lines on each side. When walking they form a kind of upright loop, whence the name "Looper caterpillars." They feed first on the young unopened buds, and, as the leaves expand, they draw two or three together with their webs, and shelter themselves within when not feeding. When full-fed, towards the end of

May (by which time they have often caused great damage), they let themselves down by a thread to the ground, bury themselves, and turn to chrysalids about two or three inches below the surface, from which the moths come up towards the end of October.

The male moths have the fore wings of an ash-grey, with various transverse markings, and the hind wings of a greyish white. The females have a most extra-ordinary appearance, from the great size of the abdomen and the small size of the abortive wings; they have no powers of flight, but fall down as if dead when alarmed, or run with some speed to hide themselves. In November the males may be seen, after sunset, flying from tree to tree, and the females creeping up the stems to deposit their eggs.—('Brit. Moths,' 'G. Chron. & Ag. Gazette,' &c.)

PREVENTION AND REMEDIES.—The best method lies in preventing the moths laying their eggs on the trees. As the females cannot fly, it is quite in our power to prevent attack to standard trees by circling the stems round, at the beginning of November, with something which they will stick fast in if they touch it, or which is too poisonous or unpleasant for them to attempt to cross.

For this purpose various mixtures are recommended, especially Stockholm tar and cart-grease, mixed in equal proportions; but, in case of the tar becoming fluid with sun-heat, it is liable to soak into the bark, and be very injurious; therefore it would appear safer, in garden cultivation, to dip a rope of rough woollen rags or a twisted hay-band in a mixture of tar and oil, which would keep moist for some time, and lay it on the ground near the tree, *all round it*, but not touching it. Probably a few spadefuls of gas-lime (such as has been exposed about a month) thrown round the stem, but not piled against it, would not be crossed by the moths. This also (if proving effectual on experiment) would meet the diffi-culty of the traffic of the moths up the posts of espaliers or up fruit-walls. The females come out about sun-down, and may be seen travelling up the stems till ten o'clock at night, so that, where the moths are numerous, a very little inspection with the aid of a lantern would show whether the remedies were acting. Also (where the moths are known to be numerous) it would be well to try shaking the boughs at night over a cloth; it is stated that when alarmed they often fall and feign death, and, if the plan answered (on trial with one tree), it would diminish the attack much at a small expense.

Trees that are infested by the caterpillars may also be cleared to a great extent by shaking. When the caterpillars are nearly full-grown they will fall to the ground, or hang by their threads on a blow being given to the bough on which they are feeding, and (as in many other cases) may be collected on a cloth spread beneath the tree and destroyed, or they may be trampled on, or quick-lime, or anything else that will kill them, thrown down on them, or they may be shaken on to tarred boards. Care should be taken to sweep down all that are dangling in the air at the ends of their threads, and also to keep them from going back up the trunk of the tree.

The means above mentioned for prevention of traffic of the moths up the stems would serve equally well in case of the caterpillars. The caterpillars bury themselves about two or three inches deep in the ground under their food-trees, where they turn to chrysalids about the end of May, from which the moths do not come up till October; so that, meanwhile, forking the surface and a good dressing of quick-lime would help to get rid of the pests; also, where it is possible to flood, or thoroughly to soak, the infested ground for a few days, it has been found a good measure of prevention.

It is desirable to remove and destroy all boughs and twigs that may be pruned late in the winter, so that there may be no chance of caterpillars hatched from eggs on the shoots or buds getting up the trees.

With regard to this pest on the large scale in which it occurs in orchard-ground, it is probable that if at the end of October a dressing was given round each tree-stem

of gas-lime that had been exposed to atmospheric·
action for a month or two, so that it would not be·
injurious to the trunk if some lay against it, that little
further trouble would be needed, and the trees would be
much benefited by the application. (For note on gas-·
lime, see references in Index.)

Encouragement of birds, and especially of Titmice and
Starlings, is serviceable with regard to this pest.

Red Spider. *Tetranychus telarius*, Linn.

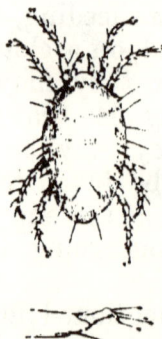

Red Spider (*T. telarius*), and foot with bristles, after Claparède.

A bad attack of "Red Spider" is a matter affecting·
the trees both for the passing season and the next ; as a
present evil the leafage, and consequently the fruit crop,
is hurt, and (from the injury to the leaves) the shoots for·
the ensuing season are also weakened.

To save trouble in reference the figs. of "Red Spider,"
Tetranychus telarius, after Claparède, and of *T. tiliarum*,
from life, are here repeated.

The egg is oval, or spherical, and colourless, and may
be found amongst the webs on the leaves. The larva
(as it is called, although the name is hardly appropriate,
the Red Spider not being an insect) hatches from the
egg in about eight days, and is much like the parent,
excepting in having only three pairs of legs ; it acquires·

the fourth pair with the change of skin (according to M. Dugès, at maturity).

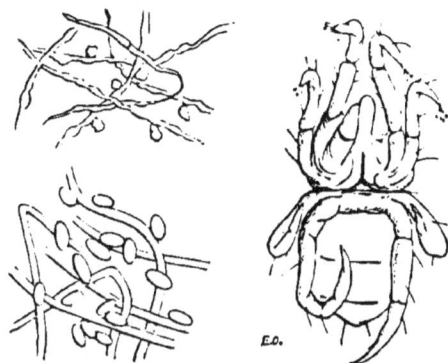

Red Spider of Lime trees (*T. telarius*, Claparède; *T. tiliarum*, Mull.); webs with eggs in dry and moist state; all much magnified.

When mature the Red Spider is oval, furnished with four pairs of legs, two pointing backwards and two forwards; the head, body, and abdomen form a solid mass, by which it is distinguishable from true spiders, which have the abdomen joined to the rest of the body by a fine stalk; and also from insects, which have the head, body (thorax), and abdomen commonly distinct from each other, and which also, in their perfect state, have never more than three pairs of legs. The head is furnished with a beak or sucker by means of which it draws the juices from the leaves, and beneath the abdomen, near the end of the tail, is a conical protuberance, from which the threads are produced with which it forms its webs.

The colour is various; of transparent yellowish white, orange, reddish, or brick-red, and other tints, depending, as far as present observations show, on the colour and nature of the food within, and partly also on the age of the individual, as these " mites " have been noted as of a green colour in early life, changing with maturity to the rust colour we are best acquainted with.

The Red Spider has difficulty in moving on perfectly smooth surfaces, but, by means of its claws and the pin-headed bristles with which they are furnished, it moves readily on the under side of the leaves, and fastens its threads to the hairs or slight prominences, thus gradually forming a coating of web, amongst which it lays its eggs, fastening them by some glutinous secretion to the threads, and amongst which also Red Spiders of all ages are to be found.

The attacked leaves may be known by their greyish or yellowish, somewhat marbled, appearance above, whilst beneath they are whitish and shiny from the covering of web. This kind of Red Spider has been found sheltered, as if for the winter, beneath stones.

Whether the *Tetranychus* of the Rose, Laurustinus, and Clover is of this species (that is, *T. telarius*), or whether these are respectively *T. rosarum*, *T. tini*, and *T. socius*, is a subject on which writers hold different opinions.—('Economic Entomology,' 'Gard. Chron.,' &c.)

PREVENTION AND REMEDIES.—The Red Spider is most injurious to vegetation in the open air, in hot dry weather; and consequently washings and syringings, or drenchings by means of the garden-engine, which will render the leafage and ground, and the walls to which the trees may be attached, moist, will be very serviceable. The extreme dryness of air and soil are thus counter-acted, and a healthy growth encouraged, which more or less counterbalances the injury to the leaves from the suction of the mites.

It is important to check the attack at the very begin-ning, and for this purpose syringings morning and evening are advised, sent hard at the under side of the leaves, so as to break the webs and wash them down with the contained mites, if possible, or more probably do good by lodging something in them offensive to the mites.

Sulphur "is the active principle and most efficient agent" in preparations for destroying Red Spider, and sulphur and soft-soap combined in various ways are amongst the most reliable remedies.

Amongst the Hop-gardens in Kent "washing the plants with soft-soap and water, or even with pure water, is a remedy for the Red Spider, and some planters tried a solution of sulphur, thrown over the plants by the ordinary washing-engines in 1868, which killed these mites."—(C. W.)

Looking at this point of sulphur being generally an ingredient in washes or applications for the destruction of Red Spider and other Acari, and the circumstance that in its crude state it does *not* combine with most of the fluids used for this purpose, may account for frequent failures in home-made applications. In order to make it combine with whatever liquid may be used, the sulphur should be boiled with an alkali, and the following recipe has been recommended : one pound of flour of sulphur and two pounds of fresh lime boiled together in four gallons of water ; or, to save the trouble of boiling, the sulphuret of lime may be purchased and used thus :—Of this sulphuret take four ounces ; soft-soap, two ounces to each gallon of hot water ; the soap and sulphuret to be well mixed before the addition of the water, which is to be gradually poured on, the mixture being stirred during the time, when a uniform fluid will be obtained without sediment, which may be used when cool enough to bear the hand, and has been found to destroy insect-pests effectually and quickly. This may be used as a syringing, or a dip for infested shoots, or well-rubbed with a brush into infested bark.

"Gishurst compound, Veitch's Chelsea blight composition, Frettingham's liquid compound, are all good. Sulphur in any form seems potent."—(A. M.)

The following recipe for dressing fruit-walls answers well as a preventive of attack :—Having obtained some soot-water, tolerably clarified and as strong a solution as can be procured, this is worked up with clay till the whole is of the consistency of thick paint, and can be applied by a common painter's brush ; to this, flour of sulphur and soft-soap are added in such proportion as may be preferred : one pound of sulphur and two ounces of soft-soap to the gallon has been found to answer. Every part of the wall is then painted with the mixture, care being taken to get it well behind the shoots, and also to paint a broad thick band along the bottom of the wall. This application, made once in the season as a regular yearly treatment, has been found to prevent Red Spider attack quite satisfactorily. — (' Gard. Chron.,' 1845.) This plan acts by poisoning and burying the "Red Spider" in the walls, and also by putting a band in the way of such as have been wintering under stones and rubbish, that they will not care to cross to get at the tree ; other mixtures, as preferred by the cultivator, might be similarly used.

Clean and properly-pointed walls are a preventive of attack, as is also ground so cultivated and attended to that there shall be no neglected surface the mites can lurk in, or hiding-places under stones, clods of earth, or rubbish beneath which they can hybernate. An autumn dressing of gas-lime would be a desirable application to neglected borders where there are infested wall-fruit trees.

If by means of experiment it should be found that there is any fluid capable of dissolving the Red Spider's webs without at the same time injuring the leafage, we could thus, by clearing off its breeding-grounds, probably get rid almost entirely of the pest.—(' Economic Entomology,' by Andrew Murray ; ' Gard. Chron. and Ag. Gazette,' Ed., &c.)

RASPBERRY.

Black Vine Weevil.   *Otiorhynchus sulcatus.*
Clay-coloured Vine Weevil.   ,,   *picipes.*
Red-legged Garden Weevil.   ,,   *tenebricosus.*

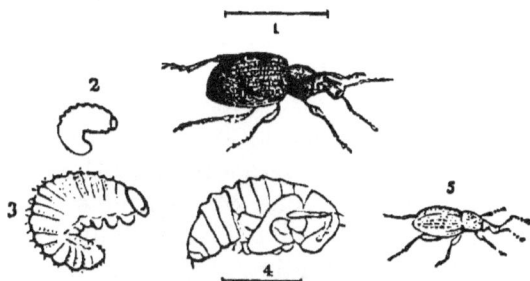

1—4, *O. sulcatus,* larva and pupa, magnified and nat. size, or with lines showing nat. length; 5, *O. picipes.*

The *O. sulcatus,* and two other kinds of *Otiorhynchus* or Weevils, are exceedingly hurtful by feeding on the shoots, leaves, and buds; sometimes also on the fruit and flower-buds; and in the larval state they are injurious by feeding on the roots.

The larvæ and pupæ of each of these three species of Weevil are very similar to the others, and so are the remedies and means of prevention; therefore I place them under one heading.

The *O. sulcatus* chiefly attacks Vines, and is also hurtful to Strawberries.

The *eggs* are laid a little below the surface of the earth. The maggots are legless, whitish, somewhat hairy, and are to be found from about August until the following spring, at the roots of their food-plants.

The pupa is yellowish white, with brownish hairs, and is to be found in April lying three or four inches below the surface, where it is stated to remain only fourteen days in this state before development.

The Weevil is of a dull black, with a short snout or proboscis; the body between the head and abdomen is

x

granulated, and the wing-cases are rough, with several raised lines; and, like the other species of *Otiorhynchus*, it has *no wings*.

The *O. picipes*, or Clay-coloured Weevil, differs from the above in being smaller and of a clay-colour, with darker spots on the wing-cases. This species is also very injurious to wall fruit-trees and Vines in hot-houses, and was identified in 1879 as being the species which caused great damage to Raspberry-plants in Cornwall.

A third species—the *O. tenebricosus*, or Red-legged Weevil—is the kind noticed as being the most hurtful to garden fruit-trees. The beetles feed on the buds, young shoots, bark, leaves, &c., of Apricots, Nectarines, Peaches, Plums, &c.; and have been found also in the maggot state doing much harm to the roots of Raspberries, Currants, Gooseberries, Strawberries, and to vegetables.

This Weevil, when recently developed, has the wing-cases dotted over with spots of delicate yellow down; these soon rub off, when the beetle appears to be of a shining black; it is sometimes of a reddish pitchy colour, whilst still immature. The wing-cases are united to each other, and the legs are bright chestnut-colour.—('Gard. Chron. and Ag. Gazette,' 'Prak. Insecten-Kunde,' &c.)

PREVENTION AND REMEDIES.—These Weevils feed by night. By day they hide away, either buried in earth by the walls against which their food-trees are trained, or round the stems of the trees, or in rough bark; crevices where mortar has fallen out of old garden walls often swarm with them, and generally they shelter in any dark nook near at hand to their nightly resort.

The commonly-adopted method of destroying these Weevils in Vineries and Peach-houses is, to spread cloths below the boughs, and shake the beetles down on them at night; then gather the beetles together and destroy them with boiling water. This plan is equally applicable to standards, and, with a little management, to wall-fruit trees; but it should be borne in mind that the beetles drop off on any disturbance, or when a light shines on them.

It is recommended that a white sheet should be laid under the boughs the day before, and a large and bright light used, so that when the beetles fall they may be easily noticed; otherwise, if no sheet has been spread and the light is dim, many of the beetles are almost certain to escape, from their colour being so like that of the ground.—(M. D.)

Another method is, for two people to hold a sheet below the boughs, and for a third to shake and then bring a light to catch the beetles by; but the above mentioned plan is better.

In 1878 the Raspberry-plots in the large fruit-gardens in Gulval and part of Madron, in Cornwall, were injured by Weevil-attack, the loss being estimated at many hundreds of pounds. On two acres of Raspberry-ground at Gulval the loss of crop from injury to the canes was estimated at one hundred pounds, and specimens forwarded from this ground in 1879 were identified as *O. picipes*.

These Weevils fed on the bushes by night, and towards dawn went down and hid themselves beneath the surface of the earth, or under stones, &c. As all the remedies tried had proved useless, a number of wooden trays were constructed, the inside of which was smeared all over with tar; and after dark one man held a tray beneath an arch (arch-training being the plan used); another, whilst carrying a lantern, gave the bush a smart tap, and the Weevils fell into the tray; the tar held them prisoners for a time, and after the tray had been placed under a bush or two, the Weevils collected were killed by pouring boiling water upon them. Thirty or forty persons were thus employed, and each bush was thus treated three times. An immense number of Weevils were caught, estimated at hundreds of thousands, and it was hoped, by continuing this plan, to avoid much future loss.—(J. T.)

On a small extent of ground the use of a common

"sweeping-net" is found to answer for clearing bushes after dark.—(W. S. M. D'U.)

Looking at the habits of the Weevil of hiding during the day in any crannies, or under clods of earth, stones, or rubbish, it would be very desirable to keep fruit-walls well pointed; also that all clods of earth, &c., should be cleared away that might serve as lurking-places. As they especially go down close to the wall, it might answer to run a line of ashes sprinkled with dilute paraffin, or with carbolic acid diluted in the proportion of one part of acid to a hundred of water, just along the junction of wall and ground.

The beetles cannot fly; therefore, as they leave the trees by day (excepting such as may lodge in the bark), any application that they could not cross would be very serviceable.

Tar has failed to prevent attack, but possibly it dried; if it was mixed with oil, so as to remain wet and sticky, and a band run along the soil at the foot of the wall, or round the stem of standard trees, it would seem impossible for the beetles to cross it.

Soft-soap well rubbed into all crannies of the bark where harbourage could take place would do good; and sulphur is of service to dust on attacked plants, but needs renewing too often to be a convenient remedy.

With regard to getting rid of the maggots:—Where attack has been going on during the growing season, the roots ought to be thoroughly examined during the winter, and all maggots that are found picked out and destroyed; and the roots dressed with hot lime, soot, or similar applications. The best remedy for an infested Vine-border (where such a plan can be carried out) is to clear it and replace it with clean soil. Watering with strong solution of ammoniacal liquor and common agricultural salt is effective in preventing the increase of this pest.—(M. D.)

Skimming off the uppermost four inches of soil for about a foot in front of the walls on which the trees were infested would be worth trying, as an experiment to be carried out further if many maggots were found; by throwing some of the removed earth into a large tub of water it would be seen directly whether there were either maggots or beetles present.

Small plants may have their roots washed out, and thus be saved; but the ground on which they grow should be thoroughly dealt with at once. A good handful of fresh gas-lime put into each hole at once is a good cure for what evil may remain.

These pests should be looked to on their very first appearance, for they are most difficult to get rid of, and severe cold has little effect on the maggots. From experiments in my own garden I found they would stand a temperature as low as 11·8°, that is, just over twenty degrees of frost, without being (as far as could be seen) in the least injured.

STRAWBERRY.

Green Rose Chafer. *Cetonia aurata*, Curtis.

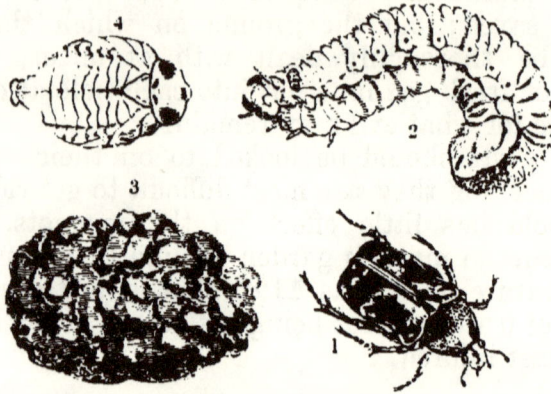

Larva; back of pupa; cocoon; and Beetle.

This Chafer is injurious both in the larval and perfect state. In the first—that is, as a grub—it feeds on the roots of Strawberries, grass, and other plants; as a beetle it frequents many kinds of flowers, including the Rose, from which it takes one of its names; but is more especially injurious by its attacks on Strawberry-blossom, and to the flowers of Turnips left for seed, where it eats off the anthers from the stamens and thus renders the flowers abortive.

The eggs are laid in the ground, where the maggots hatch and feed for two or three years. When full grown they are upwards of an inch and a half in length, thick and fleshy, of a whitish colour, with an ochreous head armed with strong jaws; the pairs of short feet are of a rusty ochreous colour, and the hinder portion of the grub or maggot is enlarged, curved towards the head, and of a lead-colour. These grubs are much like those of the Cockchafer, but are distinguished by having a horny rusty spot on each side of the segment behind the head, and by the body being clothed with transverse rows of rusty-coloured hairs; whereas the grub of the Cockchafer is almost hairless, and is without the rusty spots.

When full-fed they make earth-cases "as large as a walnut" at a considerable depth beneath the surface, which are smooth inside, but covered outside with pellets of soil which have passed through their own bodies; and in these cocoons they turn to ochre-coloured pupæ. The figure shows the appearance of the *back* of one of these, the legs and wings being folded beneath.

The Chafers, which sometimes appear as early as the beginning of May, are of the shape and size figured, of a rich metallic golden green above, with white or ochreous spots or streaks looking like cracks running across the bright green of the wing-cases; beneath they are coppery, with a rose-coloured tint. The horns are much like those of the Cockchafer, excepting that the club is formed of only three leaves (see fig. of horns of Cockchafer).

Beneath the wing-cases are large brown membranous wings, by means of which when they have finished whatever is eatable in one place they can fly with ease to another; it may be a Strawberry-bed, or may be a field of Turnips in blossom; and thus, if the weather is fine, they continue to attack whatever may attract them throughout the summer; in wet weather they die earlier in the season.—('Gard. Chron.,' ' Farm Insects,' &c.)

Prevention and Remedies.—The beetles are so large and so sluggish in dull weather that there is then no difficulty in taking them by hand; during sunshine they may be taken with a bag-net, and destroyed in any way that may be most convenient.

Where the maggots are numerous, they should be got rid of by turning over the soil, or by spreading neglected heaps of rich earth, old cucumber-beds, and similar places which they frequent, and hand-picking all that are seen. Poultry will help very much in clearing the grubs, if driven in whilst the ground is being turned over.

Where the attack is on the roots of growing plants, it

will be found serviceable to have a few tame Rooks or
Sea Gulls in the garden, as they can dig down with their
bills amongst the roots which could not otherwise be
meddled with, and clear large numbers of the grubs
without harming the plants.

The Rose Chafer when about to deposit her eggs will
sweep round on the wing until she sees a suitable spot,
and—hardly pausing from her flight—will disappear at
once down any crack that may be open in the ground, or
into a nook amongst boards, or otherwise ; and the grubs
may be found in the decayed wood-soil that accumulates
within old hollow trees.   It is therefore desirable to remove
all wood-rubbish, and also heaps of decaying sawdust
that may attract the beetle or shelter the grubs.

For other remedies, see " Cockchafer," p. 207.

LIST OF CONTRIBUTORS.*

Anderson, Alex.

Bairstow, S. D.
Boyd, Thomas.
Brown, George, Junr.
Brunton, Thomas.

Coupar, Robert.

Dewar, D.
Dickson, J.
D'Urban, W. S. M., F.L.S.
Dunn, Malcolm.

Findlay, Frank Grant.
Fitch, Edward A., F.L.S.
Fitton, Samuel.
Forbes, Alex.

Grierson, Charles.

Hart, Thomas H.

Kay, James.
Kough, S. Harley.

Loney, Peter.
Long, James.

M'Corquodale, William.
M'Donald, Thomas.

M'Gregor, John.
M'Kenzie, D. F.
M'Kinlay, George.
M'Laren, John.
Mathison, John.
Melville, D.

Norgate, Frank.

Parfitt, Edw.

Robertson, W. W.
Russell, James.

Scott, D.
Scott, D. Sym.
Service, Robert.
Shipman, W.
Shearer, Alex.
Simpson, A.
Sutherland, John.

Webster, J.
Wilkie Thomas.
Whitehead, Charles, F.L.S., F.G.S.
Whitton, J. W.

Upton, John.

* The above list contains the names of those to whom I am mainly indebted for information expressly contributed for this volume.

BIBLIOGRAPHICAL REFERENCES.

The following list contains the names of the works which have been chiefly referred to in this volume, and amongst these I wish especially to acknowledge the great amount of information obtained from the writings of the late John Curtis, contained in his 'Farm Insects,' and also in papers published in the 'Journal of the Royal Agricultural Society'; and also in the 'Gardener's Chronicle and Agricultural Gazette.'

'Manual of British Beetles.'—J. F. Stephens.
'British Beetles.'—E. C. Rye.
'Catalogue of British Hymenoptera.'—Fred. Smith.
'Die Blattwespen, und Holzwespen.'—Dr. Th. Hartig.
'Die Mitteleuropaischen Eichengallen.'—Dr. Gustav Mayr.
'Illustrations of British Entomology,'—J. F. Stephens.
'The Butterflies of Great Britain.'—Prof. J. O. Westwood.
'Introduction to Classification of Insects.'—Prof. J. O. Westwood.
'British Moths.'—Edward Newman.
'Monograph of British Aphides.'—G. B. Buckton, F.R.S.
'Die Pflanzen lause.'—C. L. Koch.
'British Hemiptera Heteroptera.'—Douglas & Scott.
'Farm Insects.'—John Curtis.
'Economic Entomology.' Aptera.—And. Murray.
'Praktische Insekten-Kunde.'—Dr. E. L. Taschenberg.
'Naturgeschichte der Schadlichen Insecten.'—Vincent Kollar.
'Journal of the Royal Agricultural Society.'
'Gardener's Chronicle and Agricultural Gazette.'
'Arboretum et Fructicetum.'—J. Loudon.
'Reports of Observations of Injurious Insects.'—Ed.

GLOSSARY.

Alate.—Winged.

Alula.—Small membranous appendage to the base of the hinder edge of the wings of Diptera (two-winged flies).

Antennæ.—"Horns" or "cranial feelers" placed in front of the head; various in form, sometimes thread-like, and longer than the insect; elbowed (see p. 234); with a club of leaves (see p. 207); also saw-like, feathery, and many other forms, in butterflies somewhat like fine small-headed pins (see p. 29).

Apterous.—Without wings.

Cauda.—Tail.

Caudal prolegs.—Sucker-feet attached to the tail-segment of many kinds of larvæ.

Chrysalis.—A term applied to the pupa or inactive stage of life of various insects, chiefly that of butterflies and moths.

Clypeus.—Portion of the front of the face above the upper lip.

Cocoon.—A case, formed of silk or other materials by various kinds of larvæ, commonly for protection whilst they change to pupæ.

Coxa.—The hip; the first division of the legs of perfect insects, between the thorax and the thigh or *femur.*

Elytra.—Horny wing-cases—term commonly applied to the upper wings of beetles.

Eyes, compound.—The common form placed on each side of the head, and composed of a large number of separate eyes placed side by side.

  „ *simple.*—Ocelli; minute single eyes placed on the crown of the head.

Feelers.—Palpi: small appendages on the lower jaws and lower lip.

Femur, plural *Femoræ.*—Thigh; second division of the leg.

Haustellum.—Term applied to different kinds of insect-mouths formed for suction.

Halteres.—Poisers; short thread-like appendages with club-shaped heads, to be found in two-winged flies, taking place of the absent hind wings.

Horns.—Antennæ; organs of various shape placed on front of head.

Imago.—Insect in perfectly-developed stage, as butterfly, fly, beetle, &c.

Labium.—Lower lip.

Labrum.—Upper lip; this is placed vertically over the lower lip, with the two pairs of jaws placed horizontally between the two lips.

Larva.—First active stage of insect life, as caterpillar, maggot, &c.

Legs.—In the perfect insect, formed of four pieces, hip, thigh, shank, and foot; in larvæ, short legs of various forms, and sucker-feet, or prolegs.

Mandibles.—Upper jaws, placed horizontally opposite each other beneath the upper lip.

Maxillæ.—Lower jaws, placed similarly to the above, just below the "mandibles," and furnished with feelers; thence called feeler-jaws.

Maxillary palpi.—Feelers on the lower jaws.

Mouth.—Formed of six pieces, an upper lip ("labrum"), lower lip ("labium"), with two pairs of jaws opposed horizontally between them, of which the upper pair ("mandibles") are often strong, and serve for biting with; the lower pair (maxillæ or "feeler-jaws") are usually smaller and weaker, and are furnished with feelers or "palpi," as is also the "labium." In Butterflies, Aphides, &c., that have sucker-mouths, some of the above-mentioned portions are variously altered in shape.—See "*Proboscis.*"

Ocelli.—Minute single eyes, usually three in number, fixed on the crown of the head, commonly known as simple eyes.

Oviparous.—Producing eggs.

Ovipositor.—Instrument whereby the female insect lays her eggs.

Ovum.—Egg.

Palpi.—Feelers, placed on the lower jaws and lower lip.

Poisers (see *Halteres*).—Appendages in the place of the hinder wings of the Diptera.

Proboscis.—In butterflies, altered lower jaws forming the spiral trunk; in two-winged flies, the altered lower lip enclosing the piercers.

Prolegs.—Sucker-feet whereby caterpillars and other larvæ hold firmly to the substance they are placed on.

Pulvilli.—Cushions beneath the feet of flies and some other insects.

Pupa.—Second stage of insect-life, in which it is often inactive, as with the "chrysalis" of butterflies.

Rostrum.—Sometimes called snout, when applied to the prolonged front of the head of some kinds of beetles (see p. 234); also used as well as proboscis for the altered labium of Aphides.

Scutellum.—A name sometimes applied to a portion of each segment, but especially to the triangular "shield-shaped" plate between base of the wing-cases in beetles and bugs.—See figs., pp. 102, 310.

Segments.—Rings of which an insect is composed—considered to be thirteen, including the head.

Spiracles.—Breathing-pores placed along sides of insect, by means of which air is drawn into or expelled from the *tracheæ* or breathing-tubes.

Stigma.—A small thickened spot on the front edge of the fore wing of various kinds of insects.

Sucker-feet.—Prolegs; fleshy appendages whereby larvæ hold firmly to the twigs, &c., whereon they are placed.

Tarsi.—Feet, the fourth piece of the leg being that on which the insect rests.

Telum.—Last segment of the abdomen.

Tibia.—Shank; third piece of the leg between the thigh ("femur") and foot ("tarsus").

Thorax.—The three segments next to the head, known respectively as the pro-, meso-, and meta-thorax, of which in perfect insects the first bears a pair of legs, the other two each a pair of legs and wings.

Tracheæ.—Organs of respiration communicating with the air by breathing-pores (spiracles) in the sides of the insect, and conveying it by means of smaller tubes throughout the body.

Ungues.—Claws or curved hooks at the extremity of the foot.

Viviparous.—Producing living young.

INDEX.

C.

D.

320

Jumpers, 106
Jumping Plant-louse, 286

L.

Lackey Moth, 261
Larch, soil and situation for, 196,
197
„ Aphis, 192
„ „ late frosts favour-
able to, 196
„ „ preventives for, 194,
195
„ blight, 192
„ blister, 198, 199
„ Bug, 192
Larva, description of, xix
Larvæ of Pear Sawfly, 289
Lasiopteryx obfuscata, 80
Leafage, healthy growth of, im-
portant, 168
Leather Jackets, 66
Lettuce Fly, 120
Lettuce-root Aphis, 119
Lime, 79, 169, 292
„ water, 32, 195
Lime-tree Mite, 204
„ Red Spider of, 204
Liquid manure, use of, 168
Lophyrus pini, 224
Lygus umbellatarum, 102

M.

Magpie Moth, 274
Mamestra brassicæ, 34
Mangold Fly, 14
Manure, liquid, 94
„ wet, to stifle grubs or
insects in, 41, 43
Marble-gall Fly, 210
„ grubs destroyed
by Titmice, 212
„ Oaks attacked by,
213
„ Oaks not observed
to be attacked by, 213
Melolontha vulgaris, 207
Mercury, perchloride of, preven-
tive for Larch Bug, 194

Mussel Scale, 266
Myzus cerasi, 272

N.

Nematus Ribesii, 277
Neuroterus lenticularis, 213
"Niggers," 163
Night-soil, 72, 253
Nitrate of soda and salt, 78
Noctua (Agrotis) exclamationis,
161
„ „ segetum, 154
„ (Tryphæna) pronuba, 37
Nut Weevil, 281

O.

Oak leaf-roller Moth, 214
„ birds of ser-
vice in clearing, 216
Odyneri, 55
Oil, caution regarding application
of, to bark, 269
Onion Fly, 123
Onions, compost and preparation
of ground for, 125, 127
„ infested, to remove, 124
„ paraffin remedy for Fly
in, 128
Orthotænia turionana, 220
Otiorhynchus picipes, 305
„ sulcatus, 305
„ tenebricosus, 305
Otter Moth, 109

P.

Pale Tussock Moth, 107
Paraffin, methods of application,
49, 50, 96, 117, 127, 128
„ dilute, a remedy for Larch
Bug, 194
„ injected into borings in
timber, 191
„ and red lead, 239
Paring and burning, 92, 117
„ disadvantages
of, 70
Paris Green, 138, 240
„ price of, 240
Parsnip Fly, 130

ERRATA.

Page 50, line 34, for D. M'L, read J. M'L.

„ 52, „ 20, „ A. G., read C. G.

„ 281, „ 2. „ *Balæninus.* read *Balaninus.*